SCIENCE AND KNOWLEDGE

Philosophical Issues, 3, 1993

PHILOSOPHICAL ISSUES
Edited by Enrique Villanueva
(Sociedad Filosófica Ibero Americana/Universidad Nacional Autónoma de México)

Previously published volumes:

CONSCIOUSNESS
(Philosophical Issues, 1, 1991)
RATIONALITY IN EPISTEMOLOGY
(Philosophical Issues, 2, 1992)

Forthcoming Volumes:

Volume 4, *Naturalism and Normativity*

Philosophical Issues, 3, 1993

SCIENCE AND KNOWLEDGE

edited by

Enrique Villanueva

SOCIEDAD FILOSÓFICA IBERO AMERICANA

Ridgeview Publishing Company • Atascadero, California

Paper text: ISBN 0-924922-14-1
Cloth (library edition): ISBN 0-924922-64-8

The typesetting was done by José Luis Olivares.

Published in the United States of America
by Ridgeview Publishing Company
Box 686
Atascadero, California 93423

Printed in the United States of America
by Thomson-Shore, Inc.

Contents

Acknowledgments

The Sociedad Filosófica Ibero Americana wants to express its gratitude to the Department of Logic and Philosophy of Science of the Universidad de Salamanca, Spain, who hosted its fourth conference on the topic of *Science and Knowledge*, June 24-26, 1991. We want to thank specially Professors Miguel Angel Quintanilla and Fernando Broncano, who together with their colleagues and members of their staff enthusiastically made possible the right environment to hold successfully this kind of meeting.

Special mention deserve, once again, Ernest Sosa and Lourdes Valdivia who were helping at each stage in the process of convening this fourth conference; without their help the conference would not have been possible. Our gratitude to both of them.

The Universidad Nacional Autónoma de México mainly through the Coordinación de Humanidades and the Dirección General de Asuntos del Personal Académico and their heads, Julio Labastida Martín del Campo and Guillermo Aguilar Sahagún, respectively, once more provided unvaluable help to make possible the celebration of this fourth conference of SOFIA.

Our gratitude for all those who participated reading papers or commenting on them and the valuable audience who support us in our aim to bring about a new reality in the philosophical work that is being done in our Iberic and Latin American countries.

I want to thank Lourdes Valdivia for helping to coordinate the volume and to José Luis Olivares for his professional and careful typesetting.

This book is dedicated to the future Ibero American Philosophers.

Enrique Villanueva
Tepoztlán, México, December 1992

PHILOSOPHICAL ISSUES, 3
Science and Knowledge, 1993

Introduction

Enrique Villanueva

The papers that compose the present volume have been arranged in three groups, first, items 1–7 dealing with issues in the Philosophy of Language; item 8 deals with a Metaphysical issue and the remaining papers deal with epistemological issues.

In *Why Meaning (Probably) Isn't Conceptual Role* Jerry Fodor and Ernie LePore throw a deep attack into Conceptual or Inferential Role Semantics (IRS). They run an argument that goes like this: Meanings cannot be identified with inferential roles for these roles are not compositional. If inferential roles are identified with roles in analytic inferences then these roles will become compositional but at the cost of implying the analytic–synthetic distinction which is an unprincipled distinction. If compositionality cannot be rejected then it seems that it is IRS with its holistic embedding that has to go thus leaving atomistic meanings and compositionality to do all the linguistic work.

Ned Block comes to the rescue of IRS in *Holism, Hyper–analitycity and Hyper–compositionality* by drawing a distinction between holism and inferential role: the former states that all inferences have to be included into the inferential roles of sentences containing a given word and that all these inferences constitute its meaning whereas, according to Block, inferential role says that only some inferences have to be included. Next, Block faces the argument that the inference from "rattling snake" to "dangerous" is analytic which is based in a principle according to which all compositional inferences become analytic. He prepares the ground for his reply by stating that what is involved in IRS are narrow contents, that is, those contents that are useful for psychological explanation. Then his reply rests in pointing that narrow contents are never analytic because they never are true (have no truth conditions) and thus are not the sort of things that could be analytic (since analytic is defined as true in virtue of the meanings involved). Block finds that the relationship between these narrow contents is better called 'constitutive' rather than 'analytic' for in this way there could be some false constitutive inferences. Thus this way constitutivity makes possible that compositionality does not entail analyticity (making false the above mentioned principle) nonetheless providing a criterion to distinguish the inferences that have to be included into the inferential roles and leaving open the possible causal characterization of these inferential roles.

In *Does an Inferential Role Semantics Rest Upon a Mistake?* Paul Boghossian deals with Fodor and LePore's argument against Inferential Role Semantics (IRS) that says that there could not be a non–holistic, compositional IRS. The argument has two premisses the first of which says that the only plausible version of a non–holistic IRS identifies an expression's meaning–constituting inferential role (MIR) with its role in analytic inference. Boghossian objects that Fodor & LePore take a simple interpretation in terms of an unreduced notion of analyticity (of valid by virtue of meaning) in its definition of MIRs.

Moreover, Boghossian grants that first premiss and directs his worries to the second premiss which says that the analytic/synthetic distinction is unprincipled. The premiss has two readings: on the first, it says that there are not any inferences(/sentences) that are valid(/true) solely by virtue of what they mean and thus that the meaning of an expression is not determined by its role in analytic inference. To this first reading Boghossian objects that one cannot be realist on meaning and reject analitycity for to accept determinate (facts on) meaning is to commit oneself to analyticity. *Contra* Block Boghossian stresses that the generality of his argument that determinate meanings imply analyticity cannot be restricted to be valid of wide contents only and not of narrow contents for even if it is expressed in terms of truth it could be recast in terms of a non–truth–conditional notion of narrow meaning. On the second reading of the premiss a much more demanding epistemic *a priori* condition is added to the purely semantical sense of anlytic so that it now says that a non–holistic IRS is committed to analyticities that are known *a priori*. This new view, Boghossian agrees, cannot be defended by a mere appeal to meaning–realism while pointing that the epistemic addendum is irrelevant and stands in need of further argument on the face of the semantical character of an IRS.

From another angle Stephen Schiffer throws doubts on compositional semantics. In *Belief Ascription and a Paradox of Meaning* Schiffer agrees that the that the best theory which assumes a compositional truth theory is the Hidden–Indexical Theory (HIT) of the logical form of belief ascriptions. Nothwithstanding the achievements of HIT, which Schiffer is prone to recognize, there remain three main problems which point to the falsity of HIT. Thus HIT comes to be both, conditionally correct and moreover false from which it follows, according to Schiffer, that natural languages lack compositional truth theories.

Besides, if the form of belief ascriptions cannot be accomodated into a compositional truth theory this has a bearing on the question about the form that a compositional meaning theory should take for a natural language. For if the HIT is correct we will know how to accomodate belief ascriptions in a compositional truth theory for English, and also will know the form a compositional meaning theory for English must take. However, a meaning paradox arises and a choice has to be made between the mutually inconsistent thesis. The choice has to be between the thesis that (1) 'That'–clauses behave syntactically like singular referential terms, (2) compositionality and (3) a compositional determination of 'that'–clause reference consistent with (1) and (2). Schiffer favours rejecting (2) because of (3), even on the face of generalized opposition by most philosophers, because he may know how to reply to two problems that arise from the rejection of (2), namely, that Mentalese (one's inner system of mental representation) must have a compositional supervenience theory which explains the systematicity and productivity of Mentalese and that these features of Mentalese explain also the systematicity and productivity of one's public language but that those explanations do not imply a compositional meaning theory unless propositions turn to be compositionally determined entities and Schiffer's view of propositions as ontologically deflationary sustains no compositionally–determined conception of the referent of 'that *S*'. Schiffer ends the paper stating that a research program in progress has to provide a solution to this paradox of meaning along these lines.

Paul Horwich in *Scientific Conceptions of Language and Their Philosophical Import* explores Quine's denunciation of the analytic–synthetic distinction and the possible use of such a distinction in cognitive science. Horwich points that the analytic–synthetic distinction rejected is a distinction that plays an epistemological role inside the reductionistic programme of logical empiricism in which each scientific hypothesis has empirical content and could be shown to have it by deducing it from facts of experience together with self–evident principles of logic. The correctness of these self–evident principles (which were conceived as stipulations that help to fix the individual meanings of their constituent expressions) is guaranteed by what they mean; they are analytic. Thus the class of meaningful hypothesis is determined. Logical empiricism also held that no sentence is immune from revision and thus whenever a sentence is revised there is a determinante fact of the matter as to whether this change does or does not imply a change of language and also a determinate fact of the matter as to whether our decision to abandon the sentence is epistemological or pragmatic.

According to Horwich, Quine rejects that there are such determinate semantic facts of the matter for we can explain both belief–change and scientific respectability of sentences within a holistic model without individual empirical contents. And if linguists or psychologists wanted to introduce a technique to determine analytic sentences that ability will not be useful for science or for philosophy of science. Meaning, according to this reading of Quine, is heavily indeterminate and serves no useful theoretical purpose.

Horwich sees a way to reconcile Quine's rejection of meaning with the advocates of a scientific theory of language (like Chomsky) who find no such indetermination in meaning or in the semantic properties; this is found when it is realized that Quine is addressing the common or ordinary notion of meaning whereas the linguists have introduced a new use, new rules, and have framed a new notion of meaning, a scientific one. Moreover, Horwich concedes, first, that he has no proof that any scientific concept of meaning has to differ from the ordinary concept; and second, that Quine's philosophical important critique lies not in neglecting the existence of analyticity or the framing of a new technical notion of analyticity but rather in the epistemological irrelevance of the ordinary notion of meaning. To end, Horwich extracts the following moral: if philosophy proves to be irrelevant to science, scientific results will not solve philosophical problems.

Paul Boghossian rejects both the distinction between an ordinary behaviouristic folk concept of meaning and a new cognitive concept of meaning and the idea that analyticities only spring from the last, cognitive concept. In *Cognitive Science and the Analytic/Synthetic Distinction: Comments on Horwich* Boghosian argues against Horwich way of making compatible Quine's rejection of the analytic/synthetic distinction with the acceptance by cognitivist scientists' of that distinction because it relies on Quine's behaviourism on meaning —the idea that behaviorally constituted meanings are indeterminate— which is all wrong and must be rejected. Boghossian finds Horwich's arguments that verificationism cannot characterize the notions of synonymy and analyticity and that the holism of the web–of–belief model avoids notions of analyticity unable to prove the indeterminacy of those behaviouristically constituted meanings.

But even granting the above mentioned arguments Boghossian thinks that Horwich cannot make the step that goes from the meanings behaviourally constituted to the meanings attributed in folk psychology where the character and nature of the mind/brain is what underlie linguistic activity. Wittgensteinian arguments to the contrary, two points witness the conclusion that public–language meaning is not behaviouristic, namely, on the one hand, that an account of the epistemology of other minds could be given by using a mentalistic analogically mediated inference (to which the behaviourist himself has to make recorse), and on the other, that there is an indeterminacy from the observable behaviour to the mental (meaning) property ascribed. Accordingly, Boghossian claims to have made room for a different picture, one in which there are no two notions of meaning but only one which is mentalistic, determinate, supports an analytic/synthetic distinction and is related to the empirical study of meaning.

In *On the Logical Form of Propositions: Some Problems of Vanderveken's New Theory of Propositions* Jorge Rodríguez Marqueze presents two difficulties to the view of propositions held by Daniel Vanderveken: the first concerns the structure of second or higher order propositions which holds Rodríguez Marqueze is not the one needed to reflect what a speaker must have in mind when he expresses that proposition with an illocutionary force in an utterance; Vanderveken's analysis of these higher order propositions commits him both to the acceptance of direct reference for all singular expressions referring to propositions and to the admission of singular propositions and those two are incompatible with his analysis of propositions. The second difficulty, again *ad hominem*, is that when Vanderveken's analysis of propositions is applied to propositions expressed by sentences containing indexical expressions it turns out to be inconsistent with his principle of substitutivity for occurrences of sentences in contexts, leaving no room for non–synonymous indexical expressions with the same semantic values, claims Rodríguez Marqueze.

Ernest Sosa goes into a regal topic arguing that the central concepts of epistemology derive from that of truth. In *The Truth of Modest Realism* he starts dialectically in order to motivate his own position. He rejects neo–pragmatist as well as neo–peircean accounts of truth thus resting with a correspondence account. *Contra* Hilary Putnam he rejects that perspectival truth implies that reality has got to be "internal" to a perspective. On the contrary, Sosa holds that the perspectival nature of truth requires that reality remains absolute and the source of perspectival truth and falsity.

Moreover, Sosa agrees that we lack a philosophically substantive account of truth, namely, one that receives a substantive correspondence analysis in which the reference of sentential constituents will be crucially involved for such reference will work through some sort of appropriate causal relations. Lacking this much we should opt for a deflationary account which could be a substitutional paraphrase account (like that of Michael Williams) or a finitely minimal theoryof truth (as Paul Horwich claims). This deflationary account serves the main purpose of letting us to appeal to truth when we explain our individual and collective success claiming that enough of our beliefs and theories have it.

A deflationary account is what we could have now, that is, an account of the sort that Moore and Russell held at some early stage in their careers, one such that truth is conceived as a primitive, simple, unanalyzable property that is possesed by some propositions but not by others. Such austere notion does not give an illuminating, compact, surveyable analysis that could provide a philosophical explanation of truth. This austere notion which postulates propositions, facts, an expression relation and a sort of streamlined and abstract "correspondence" with the facts we need for it applies across languages, is antecedent of the language of any given speaker; we need it as a central notion for epistemology. But acceptance of this primitivism in the case of truth does not preclude a philosophical theory of such truth that might turn to be perspectival in character for it does remain as an open possibility for Modest Realism, according to Sosa.

In *Mental Causation in a Physical World* Jaegwon Kim points out that Descartes' form of dualism could not cope with mental causation and the pairing problem i.e. how do we know that the right cause has been paired with the right effect? represents an irresolvable metaphysical difficulty for Descartes' view. This bifurcated model was replaced with a layered world model, one in which a hierarchical stratified system of "levels", or "orders", of entities and their characteristic properties in which entities belonging to a given layer are mereologically composed of entities belonging to the lower levels. A central question rised, namely, are properties associated with different levels also ordered by some distinctive and significant relationship? Kim holds that the major rival positions addressing the mind body problem can be seen as answers to that central question.

Microdeterminism i.e. the view that once all the facts about the basic particles of the world are fixed at any instant of time, that fixes all the facts about the world at that instant, transforms the problem of mental causation into the problem of finding a causal role for mental events in this microdeterministic world, that is, Is mental causation possible in this microdeterministic world? Most physicalists are not reductionists but Kim argues that there is no nonreductivist explanation of mental causation establishing that all mental causation goes through mental–physical causation and given the principle that states that there can be no excess causal powers accruing to higher properties that are not also among the powers of lower properties on which the higher properties depend, no additional causal work is left for mental properties to carry on.

Type reductionism takes mental causation as a special case of physical causation and tries to solve it by this route but the fact of multiple realization i.e. that the nomologically possible physical bases of a mental property are so diverse and heterogeneous that they cannot be thought to constitute a unitary physical property (for disjunctive properties of this kind fail to be nomic properties) that could serve as the reduction base for the mental property should be taken as implying that mental properties are not well behaved, appropriately nomic and fit for scientific theory.

From the above Kim goes on to draw a broad philosophical position in which we lose mental properties as nomic properties (even if these properties could have some other uses and enter some specific causal roles) and no reduction task is left. Moreover, mental concepts remain (may be as indefinable and functionalism might be true of the meanings of mental expressions) and they might be used to describe the instantiation of a physical property in an appropriate setting because the causal powers of the mental property have been reduced, multiply and disjunctively to the causal powers of its physical realizers. Kim concludes that this is all the reduction we need and want for multiply realized properties.

Richard Jeffrey calls attention to Jon Dorling's Bayesian rebuttal of the skeptical conclusion often drawn from the presumed arbitrariness of answers to the question of which to give up —theory or auxiliary hypothesis?— when they jointly contradict empirical data. In *Taking back the day! Jon Dorling's Bayesian Solution to the Duhem Problem* Jeffrey examines Dorling's achievements to supplement deductive logic with Bayesian methodology in order to get rid of the skepticism that claims that "scientific methodology is nothing but politics" and then go on to mount a full scale attack on the rationality of men of science and furthemore go on to look for purely sociological explanations for cases of changes in scientists' beliefs.

Bayesian analysis, claims Jeffrey, vindicates the rationality of experienced scientists' reactions to evidence by supplementing standard logic with an account of the logical relations between degrees of belief which fall short of certainty. Subjective probability theory provides such an account — as Jeffrey works to show in four different cases— and is the simplest account we posess for it provides standards for the validity of informal deductive reasoning comparable to those which traditional logic has provided for the validity of that same informal deductive reasoning. Jeffrey claims that these standards can be provided with a rationale and justification quite independent of any appeal to the actual practice of scientists, or to the past success of such practices.

In *What is to be said for simplicity?* Richard Foley explores the consequences of the lack of epistemic reasons —reasons implying truth— to believe that the simplicity of an hypothesis should make it a rational choice. Foley notices that it is very difficult to find a rationale for thinking that the simpler of two hypothesis, all else being equal, is more likely to be true. On the face of failure of verosimilitude or approximate truth as the simplest explanation of the increasing predictive power of physical sciences, he opts for assuming that there is no such rationale and thus that simplicity is not a mark of truth.

Once the above assumption is made it does not follow that accounts of rational belief that emphazise considerations of simplicity are inadequate but rather that it is best not to interpret them as accounts of epistemic reasons but as being concerned with what we have reasons to believe insofar we have a broader range of aims. That means adopting an attitude weaker than belief like that of being committed to the truth of the hypothesis. Foley draws a distinction between committment and belief and argues that there is no positive rationale for the disposition of thinking that believing simple hypothesis is preferable to a policy of committing ourselves to them. Such a disposition lacks adequate reasons and thus is irrational but with a irrationality of a very weak kind: a weakness that lacks reasons to get rid of that sort of disposition, and on the contrary, one that might even be positively rational for us to have. It may be rational for us to be irrational in this way, thus concludes Foley.

A challenge of the tacit theory that posits an internally represented knowledge structure —a body of rules or principles or propositions— which serves to guide the execution of the capacity to be explained is presented by the simulation theory. In *Folk Psychology: Simulation or Tacit Theory?* Stephen Stich and Shaun Nichols submit the simulation theory to close scrutiny to conclude that it does not make good its claims to both eliminate the tacit theory and validate the simulation view. The simulation theory says that instead of an internally represented folk psychological theory plays a central role in the excercise to our ability to describe, predict and explain people's behaviour there is a special sort of mental simulation in which we use ourselves as a model for the person we are describing or predicting and in the cognitive mechanisms subserving these abilities.

Stich and Nichols notice that the simulation theory not only eliminates the internally represented knowledge structures but also that the folk psychology which underlies our ordinary explanatory practice is not a theory at all. The challenge being such Stich and Nichols proceed to examine the way simulation might be used to predict and explain behaviour and on producing intentional descriptions or interpretations and furthemore on explaining the meaning of intentional terms like 'believes' and 'desires'. Stich and Nichols find that simulation theory cannot get validated either by dismissal of the tacit theory or by proving that there are simulation accounts of abilities as grammar acquisition of abilities to anticipate the behaviour of middle sized physical objects. On some other counts evidence allows the survival of both competing theories. Finally, Stich and Nichols present developmental data and the cognitive penetrability of our predictions and explanations of behaviour as facts that are difficult to accomodate by the simulation theory but not so by the tacit theory. The prospects in favour of the simulation theory thus are dim and this being so cognitive scientists and philosophers cannot take it seriously, conclude Stich and Nichols.

In *Epistemic Folkways and Scientific Epistemology* Alvin Goldman develops a reliabilist account departing in a crucial respect to spouse an approach in terms of intellectual virtue. Goldman deals with the descriptive part of scientific epistemology in order to explain the concept of justified belief for which he has to describe the psychology of the epistemic evaluator who posses and uses that concept of justification. Goldman provides such account in two stages, namely, in the first the evaluator gets a list of virtues and vices that comes from the reliability or judged propensity he ascribes to each process to produce a high ratio of true beliefs. In the second stage the evaluator applies his list of virtues and vices to decide the epistemic status of targeted beliefs.

Goldman thinks that his veriety of reliabilism is preferable to other types because on the one hand, it solves the puzzle cases presented to any reliabilist account (cases like the demon–world, clarivoyance or disease–triggered cases), and on the other hand, it has not to relativize the virtue or vice to a relevant domain or to a field and environment. In this last respect Goldman argues that even if it is needed an explanatory story of how the folk arrive at their selected lists of virtues and vices, we find no evidence of relativization when describing the epistemic folkways. Similarly, an account is needed of the principled methods for individuating cognitive processes or for slicing up virtues and vices but it is not the descriptive part of scientific epistemology the one that has to fulfil that need.

Intellectual virtues according to Goldman are faculties or processes that are accepted as answers to questions like "How does he know?" to which answers like "because he saw it" qualify as right and thus become virtues and on the opposite, answers like "because he guessed" are anomalous and are rejected and become vices. To end, Goldman notices analogies with rule utilitarianism in the moral sphere and with reference–fixing in semantics where once a referent has been determined, it tends to retain that status in future without being reassessed each time a new case is considered and without minding the way used to get to that very reference neither having to relativize it to each possible world or environment.

Josep Corbí addresses Goldman's theory in *In Defense of a Normative Scientific Epistemology* in order to ground his view that a scientific epistemology must remain continuous with our folk epistemic practice because of rational constrains derived from interpretation theory; that it must be normative in the sense of a reliabilist criterion of rightness because a similarity constraint that an interpretation theory extends to all believing subjects and because a principle of simultaneity stipulates that the description of rule–governed and epistemic practices cannot be carried out without assessing them at the same time. And finally, that a scientific epistemology must be scientifically informed both (1) in the sense that it is a scientific task to elucidate and describe the ordinary epistemic concepts and (2) that the results of cognitive science are relevant to determine the basic belief–forming processes to which scientific epistemology applies the normative standards detected in (1) on the condition that intentional psychology is incorporated in cognitive science.

In *Epistemic Aims and Values in W.V.Quine's Naturalized Epistemology* Javier Rodríguez Alcázar asks if the cognitive aims and values are outside science or are to be approached from within, using sience's own methods. Rodríguez Alcázar stirs into quinean orthodoxy to find the strains caused when aims and values are included into the scientific task itself in Quine's Naturalized Epistemology's View and declares himself in favour of a radicalization of Quine's naturalism making these values and aims internal to science on pain of falling into incoherence.

1

PHILOSOPHICAL ISSUES, 3
Science and Knowledge, 1993

Why Meaning (Probably) Isn't Conceptual Role[1]

J.A. Fodor and E. LePore

Introduction

It's an achievement of the last couple of decades that people who work in linguistic semantics and people who work in the philosophy of language have arrived at a friendly de facto agreement as to their respective job descriptions. The terms of this agreement are that the semanticists do the work and the philosophers do the worrying. The semanticists try to construct actual theories of meaning (or truth theories, or model theories, or whatever) for one or another kind of expression in one or another natural language; for example, they try to figure out how the temperature could be rising compatible with the substitutivity of identicals. The philosophers, by contrast, keep an eye on the large, foundational issues, such as: what's the relation between sense and de-

[1]The material in this paper is adapted from our forthcoming book *Holism: A Shopper's Guide*, Basil Blackwell, 1991.

notation; what's the relation between thought and language; whether translation is determinate; and whether life is like a fountain. Every now and then the philosophers and the semanticists are supposed to get together and compare notes on their respective progress. Or lack thereof.

Accordingly, this paper is not about semantics but about the philosophy of language. Jerry and I have been poking around in the basement of meaning theory, and we seem to have discovered a large, nasty crack; as far as we can tell, one of the foundation stones is coming unstuck. We thought we'd better tell you about it before things get worse.

We'll proceed as follows: First, we'll try to say where in the foundations the problem is located; then we'll try to say what the problem is; and then we'll make a suggestion or two about what to do about the problem. The first part of the discussion will be very broad; the second part will be rather more specific; the third part will be practically nonexistent.

Here goes.

1 Where the problem is

A traditional foundational problem in the theory of meaning is: *Where do semantic properties come from?* The presupposition of this question is that the fact that a word (or a sentence, or whatever) means what it does can't be a brute fact. It can't be a brute fact, for example, that 'dog' means *dog* and not *proton* and that 'proton' means *proton* and not *dog*. Rather, 'dog' must have some *non*–semantic property in virtue of which it means *dog* and not *proton*; and 'proton' must have some (different) non–semantic property in virtue of which it means *proton* and not *dog*. To put it in the standard philosophical jargon, semantic properties must *supervene* on non–semantic properties. There may be some properties that things *just* have; that they have for no reason at all. But if there are, they are the kinds of properties that basic physics talks about (like mass, charge, and charm). They certainly don't include the kinds of properties that semanticists talk about (like meaning *dog* or being a synonym of 'bachelor').

We remark in passing that none of this is to be construed as an attempt to legislate physicalism. For present purposes we are content that semantic properties should be, for example, irreducibly intentional, or irreducibly epistemological, or irreducibly teleological. But we take it to be not in the cards that they are irreducibly semantic. In short, we don't care whether semantic properties supervene on something that is *physical* just as long as they supervene on something other than themselves.

So the question arises: What *do* the semantic properties of symbols supervene on? Over the years, philosophers of language have been enthusiastic for two quite different (perhaps, indeed, incompatible) kinds of answers to this question. There's what we will call the 'Old Testament' story, according to which the meaning of an expression supervenes on *the expression's relation to things in the world*; and there's what we will call the 'New Testament' story, according to which the meaning of an expression supervenes on the expression's *role in a language*. The disagreement between these two sorts of story is venerable, and we don't propose to go into the details here. Just a paragraph or two by way of assembling reminders.

Old Testament semantics derives most directly from the British Empiricists and has, among its modern representatives, behaviorist psychologists like Watson and Skinner and a handful of philosophers in the 'naturalized semantics' movement (including Dretske, Stampe, Millikan, Papineau, Barwise and Perry in certain of their moods, and one or more of the Fodors). The basic idea is that 'dog' means *dog* because of some (non–semantic) relation that holds between the symbol and the animal. If you are a British Empiricist (or, at least, if you're Hume), you say that this relation comes down to some variant on resemblance; 'dog' means *dog* rather than *proton* because 'dog' is associated with a certain mental image that means *dog* rather than *proton*. And the mental image means *dog* rather than *proton* because it resembles dogs quite a lot but resembles protons hardly at all. If you are a behaviorist psychologist, you say that the symbol–world relation that semantic properties supervene on is *causal* (typically associative); 'dog' means *dog* rather than *proton* because, in consequence of the speaker's history of conditioning (or something) dogs cause utterances of 'dog' and protons don't. Recent developments of Old Testament Semantics propose still other variations on this theme, including, for example, appeals to nomological and/or informational relations between symbols and the world.

Now, this Old Testament story about meaning has come under a lot of criticism, both in philosophy and in linguistics. Perhaps the basic objection is one that derives from Frege: Meaning *can't* be a symbol–world relation, according to this objection, because *identity* of symbol–world relations is compatible with *difference* of meaning. Thus, it's plausible that the expressions 'the Morning Star' and 'the Evening Star' are both attached to the same non–linguistic thing (viz. to Venus); but it's also plausible that they don't *mean* the same thing. If they did, you couldn't deny that the Morning Star is the Evening Star without self contradiction or assert that it is without tautology. Both of which, in fact, you can.

So, then, according to this argument, the meaning of an expression doesn't supervene on the way that it's attached to the world. So what does it supervene on?

The New Testament story is an elaboration of the following idea: the expressions 'the Morning Star' and 'the Evening Star' mean different things, despite their both being attached to Venus, because they have different *roles in the (English) language*. Frege says "only in the context of a sentence does a word have a meaning" and Wittgenstein adds that "to understand a sentence is to understand a language". The core idea is that it's the way they are connected *to one another* that determines what the expressions in a language mean. Very often (in fact, in all the semantic theories that we will discuss) this notion of 'role in the language' is given an epistemological twist: To master a linguistic expression is to recognize the validity of some core set of *inferences* that fix its semantic relation to other expressions in the language. To master the expression 'dog' is, inter alia, to recognize the validity of inferring 'Rover is an animal' from 'Rover is an dog'. To master the expression 'The Morning Star' is, inter alia, to recognize the validity of inferring from 'is the Morning Star' to 'rises in the Morning'. And so forth.

We want to emphasize that, at this stage, we're using notions like *infer* and *disposed to infer* as blank checks. For the moment we'll let any disposition to have one belief cause another count as a disposition to infer the second from the first. So deductive inferences, and inductive inferences, and plausible inferences, and prudential inferences, and mere associations, and Heaven knows what all else, are included pro tem. A lot of this discussion will be about the problems that arise when a New Testament Semanticist tries to say exactly which inferences are constitutive of the meanings of the terms that enter into them.

This New Testament picture is quite close to one that's familiar from Structuralist linguistics, according to which the meaning of an expression is its role in a 'system of differences'. To know what 'dog' means is to know that it excludes 'cat' and cohabits with 'animal'. To know what 'bachelor' means is to know that it excludes 'spinster' and cohabits with 'unmarried'. Notice that it follows from this view that if English didn't contain words that mean what 'cat' and 'animal' do, *it couldn't contain a word that means what 'dog' does.* This is a kind of conclusion that structuralists in semantics have, with varying degrees of enthusiasm, quite generally been prepared to embrace.

Although, as far as we can tell, one or other version of the New Testament view is held practically as gospel not only in linguistics and philosophy, but throughout the Cognitive Sciences, it is nevertheless possessed of well known problems. For example, New Testament Semantics appears to be intractably holistic. This is because, once you start identifying the content of a belief with its inferential role in a system of beliefs (mutatis mutandis the meaning of an expression with its inferential role in a language), it's hard to avoid proceeding to identify the content of a belief with its *whole* inferential role in the systems of beliefs. And, having gone that far, it's hard to avoid the conclusion that if belief systems differ at all in respect of the propositions they endorse, then they differ entirely in respect of the propositions they can express.[2] This is a well–greased, and well travelled, slippery slope; having

[2] The issues we'll be concerned with arise in precisely parallel ways for

arrived at the bottom, one finds oneself accepting such prima facie outlandish doctrines as that there is no such relation as translation; that no two people ever mean the same thing by what they say; that no two time slices of *the same* person ever mean the same thing by what they say; that no one can ever change his mind; that no statements, or beliefs, can ever be contradicted (to say nothing of refuted); and so forth. It's a moot question how to get the goodness out of inferential role semantics without paying this extravagant price. Indeed, it's moot whether it is possible to do so.

Serious though these worries are, however, they are all, as it were, *external* to functional role semantics as such; they impugn its consequences for epistemology and ontology rather than its coherence. And a convinced New Testament Semanticist might be prepared to bite the bullet. If the doctrine that meaning supervenes on intra–linguistic relations has relativistic, idealistic or even solipsistic consequences, then perhaps we had better learn to live with relativism, idealism or even solipsism. There are many, especially on the West Coast, who clearly long to do so.

Well, so much for the background; now here's the problem. It looks as though there are serious *linguistic* problems for the conceptual role approach to meaning. We want to argue that, quite aside from ontology and epistemology, it looks like New Testament Semantics can't be squared with a pair of linguistic hypotheses that practically everybody thinks there are very good reasons to endorse; and which, de facto, most New Testament Semanticists actually *do* endorse. These are, first, that natural languages are compositional, and, second, that the a/s distinction is unprincipled. (This means not that the distinction is *vague* —what empirical distinction isn't?—, but that there aren't any expressions that are true or false solely in virtue of what they mean.) The foundational problem to which we wish to call your attention is that these two principles (compositionality and the denial of the a/s distinction) together with New Testament Semantics form

theories of the meanings of linguistic expressions and theories of contents of thoughts. In the rest of what follows, we will not bother to draw this distinction unless something turns on it.

an inconsistent triad. At least one of the three will have to go.

The next section of this paper sketches our argument for this claim; the third asks what on earth to do about it.

2 Compositionality, analyticity and inferential role semantics

There is, we believe, an internal connection between analyticity and compositionality that hasn't previously been remarked upon and that has dire implications for the New Testament identification of meanings with inferential roles. First, a few remarks about compositionality.

A language is *compositional* iff (idioms aside) the meaning of its syntactically complex expressions is a function of their syntactic structures together with the meanings of their syntactic constituents. For present purposes, a language is compositional iff the meaning of its sentences is a function of their structural descriptions together with the meanings of their lexical constituents. We take the doctrine that natural languages are compositional to be, as one says in Britain, non–negotiable.

No doubt connectionist psychologists (and a handful of philosophers) have occasionally been moved to dispense with compositionality. But that just indicates the desperateness of their plight. For, compositionality is at the heart of some of the most striking properties that natural languages exhibit. The most obvious of these are *productivity* (roughly, the fact that every natural language can express an open ended set of propositions) and *systematicity* (roughly, the fact that any natural language that can express the proposition P will also be able to express many propositions that are semantically close to P. If, for example, a language can express the proposition that aRb, then it can express the proposition that bRa; if it can express the proposition that $P \to Q$, then it can express the proposition that $Q \to P$; and so forth.)

We digress to remark (since the issue often comes up) that pairs like 'John calculated the answer' and *'The answer calculated John' are *not* exceptions to the systematicity of English. Systematicity requires that if a language can express the proposition P and the proposition P is semantically close to the proposition Q, then the language can also express Q. If, however, there *is no* such proposition as Q, it is no objection to its systematicity that a language can't express it. We are inclined to assume that there is no such proposition as that the answer calculated John; hence, that the present examples do not constitute an objection to the systematicity of English. We do not, however, wish to dogmatize. If there *is* such a proposition as that the answer calculated John, then English *can* express it; indeed the form of words "the answer calculated John" does so. English is thus systematic on either assumption.

Connected to both productivity and systematicity is a further, apparently perfectly universal, feature of natural languages. The structure of sentences is, in the following sense, *isomorphic* to the structure of the propositions they express: *If a sentence S expresses the proposition that P, then syntactic constituents of S express the constituents of P.* If, for example, a sentence expresses the proposition that P and Q, then there will be one syntactic constituent of the sentence that expresses the proposition that P and another syntactic constituent that expresses the proposition that Q. If a sentence expresses the proposition that John loves Mary, then there will be a syntactic constituent of the sentence that refers to John, another syntactic constituent of the sentence that refers to Mary, and another syntactic constituent of the sentence that expresses a relation such that, necessarily, that relation holds between x and y iff x loves y. Notice that, though all of this is patent, none of it is truistic. Idioms and other 'holophrastic' constructions are all exceptions, albeit the sorts of exceptions that prove the rule.[3]

---

[3]Our statement of the isomorphism principles is intentionally left pretty vague; deep issues arise when more precision is attempted. Suppose, for example, that you hold that (in a null discourse) the sentence 'It's raining' expresses the proposition that it's raining here. Then either you must say that 'it's raining' has more constituents than appear on its surface or that the isomorphism principle can be violated by prag-

Our point is that these three generalizations about natural languages —productivity, systematicity and isomorphism— are connected and explained on the assumption of compositionality; they all depend upon the principle that the meaning of a sentence is composed from the meanings of its parts. And they appear to be completely baffling otherwise. So, we intend to insist on compositionality in what follows.

Now we get to the crack in the foundation. It turns out that compositionality is an *embarrassment* for the kind of New Testament Semantics that identifies the meaning of an expression with its inferential role. In particular, it invites the following kind of prima facie argument:

— Meanings are compositional.

— But inferential roles are *not* compositional.

— So, meanings can't be inferential roles.

The second step is, of course, the one that's doing the work. But it seems pretty obviously sound. Consider the meaning of the phrase 'brown cow'; it depends on the meanings of 'brown' and 'cow' together with its syntax, just as compositionality requires. To a first approximation, 'brown' means —if you like, it connotes the property— BROWN, 'cow' means COW, and the semantic interpretation of the syntactic structure $(\text{ADJECTIVE} + \text{NOUN})_N$ is property conjunction. (We are aware that there are problems about decoy ducks and the like; but the assumption that language is compositional is the assumption that such problems can be solved.) But now, prima facie, the inferential role of 'brown cow' depends not only on the inferential role of 'brown' and the inferential role of 'cow', *but also on what you happen to believe about brown cows.* So, unlike meaning, inferential role is, in the general case, not compositional.

Suppose, for example, you happen to think that brown cows are dangerous; then it's part of the inferential role of 'brown cow' in your dialect that it does (or can) figure in

matically carried information. For present purposes, we propose not to broach these sorts of issues.

inferences like '*brown cow → dangerous*'. But, first blush anyhow, this fact about the inferential role of 'brown cow' doesn't seem to derive from corresponding facts about the inferential roles of its constituents. You can see this by contrasting the present case with, for example, the validity of inferences like '*brown cow → brown animal*' or '*brown cow → not–green cow*'. 'Brown cow' entails 'brown animal' because 'cow' entails 'animal'; 'brown cow' entails 'not–green cow' because 'brown' entails 'not–green'. But it doesn't look like either 'brown' or 'cow' entails 'dangerous', so, to this extent, it doesn't look like the inference from 'brown cow' to 'dangerous' is compositional.

In short, it appears that some, but not all, of the inferential potential of 'brown cow' (some of its 'role in the language') is determined by the respective inferential potentials of 'brown' and 'cow', the rest being determined by ones "real world" beliefs about brown cows. This should not seem surprising or contentious; it's just a way of saying that 'brown cows are dangerous' (unlike 'brown cows are animals' and 'brown cows are not–green' and 'brown cows are brown') is clearly synthetic; i.e., 'brown cows are dangerous' is *contingently* true, true in virtue of the facts about brown cows, not in virtue of the facts about meanings (assuming that it's true at all).

But, to repeat, if meanings are compositional and inferential roles aren't, then it follows that meanings can't be inferential roles.

As far as we can tell, this line of argument is quite robust; in particular, it doesn't depend on detailed assumptions about how an inferential role semantics construes the notion of inferential role. There is, for example, an influential paper by Hartry Field in which inferential role is analyzed in terms of subjective probabilities; in effect, the inferential role of your thought that P is identified with the subjective probability that you (would) assign to P contingent on each of the other thoughts that you (can) entertain (Field, 1977). So, for example, the inferential role of your thought that it's raining is determined in part by the subjective probability that you (would) assign to that thought on the assumption that the streets are wet; and in part by the subjective probability that you (would) assign to it on the assumption that the sun is out; and in part by the subjective probability that you (would) assign to it on the assumption that elephants have wings...and so forth. Like other species of New Testament Semantics, this treatment generalizes, in fairly obvious ways, from a theory according to which inferential roles are assigned to thoughts to one in which they are assigned to linguistic expressions, or to both.

Our point is that the construal of inferential roles in terms of subjective probabilities, whatever other virtues it may have, does nothing to help with the compositionality problem. This is because *subjective probabilities are not themselves compositional.* For example, the subjective probability one assigns to the thought that (*brown cows are dangerous*) is not a function of the subjective probability one assigns to the thought that (*cows are dangerous*), together with the subjective probability that one assigns to the thought that (*brown things are dangerous*). If this seems unobvious, consider a world (or rather a belief world, since the probabilities at issue are supposed to be subjective) where there are very many things that are cows, almost none of which is dangerous, and very many things that are brown, almost none of which is dangerous, and a very small number of brown cows, almost all of which are very, very fierce. On these assumptions, the probability that something that is brown is dangerous is small, and the probability that something that is a cow is dangerous is small, but the probability that a brown cow is dangerous is as big as you please.

We intend the argument so far as one horn of a dilemma, and we anticipate the following reply: "OK, so if the compositionality of meaning is assumed, meanings can't be identified with inferential roles as such". But this doesn't *really* embarrass New Testament Semantics because meanings can still be identified with roles in *analytic* inferences. Thus, on the one hand, the inference 'brown cow → brown animal' is compositional (it's inherited from the inference 'cow → animal'); and, on the other hand, precisely because it is compositional, 'brown cow → brown animal' is analytic. Compositional inferences will always be analytic and analytic inferences will always be compositional; the compositionality of an inference and its the analyticity of an inference are the same thing.

Look at it this way: If 'brown cow → brown animal' is compositional, then its warranted by the inferential roles of the expressions 'brown' and 'cow'. That's what it is for an inference to be compositional. But, according to New Testament Semantics, the inferential roles of 'brown' and 'cow' *are their meanings.* So, then, that 'brown cow → brown animal' is warranted follows from the *meanings* of 'brown' and 'cow': But for an inference to be analytic *just* is for it to be warranted by the meanings of its constituent expressions. So the compositionality of 'brown cow → brown animal' —or, mutatis mutandis of any other inference— *entails* its analyticity. The same argument also works the other way around: for an inference to be analytic is for it to be warranted by the meanings of its constituents. But, according to New Testament Semantics, meanings are inferential roles. So, for an inference to be analytic is for its warrant to be determined by the inferential roles of its constituents. But for the warrant of an inference to be determined by the inferential roles of its constituents is for the inference to be compositional. So compositionality entails analyticity and vice versa. So, then, meaning is compositional and inferential role isn't and role in analytic inference is. What all that shows is just that we need a revised version of New Testament Semantics; one which identifies meaning with role in analytic inference.

The first thing to say about this new suggestion is that the threat of circularity is now very close to the surface. It is proposed that we reconcile CRT with the compositionality of meaning by identifying the meaning of an expression with its role in analytic inferences. But the difference between analytic inferences and inferences *tout court* is just that the validity of the former is guaranteed *by the meanings* of their constituent expressions. So analyticity, meaning (and compositionality) scrape out a living by doing one another's wash, and Quine gets to say 'I told you so'.

Notice also that the naturalizability of inferential role semantics is jeopardized by the present proposal. A lot of the attraction of identifying meaning with inferential role lies in the thought that the inferential role of an expression might in turn be identified with *causal* role, thereby conceivably providing the basis for a naturalistic solution to Brentano's problem. That causal relations reconstruct inferential relations is a foundational assumption of computational theories of mental processes, so perhaps there is hope here of a unification of semantics with psychology. But, barring proposals for a causal theory of analyticity, this tactic in unavailable to the philosopher who identifies meaning with the role of an expression in *analytic* inference.[4] The idea that mental processes are computational may provide the basis for a naturalistic account of inference, but it offers no insight at all into the nature of analyticity. Nor, as far as we can see, does anything else.

We can now say pretty exactly what our problem is: You can't identify meanings with inferential roles tout court, since, unlike meanings, inferential roles tout court aren't compositional. You *can* identify meanings with roles in *analytic* inferences, however, because analytic inferences *are* compositional. But, of course, the cost of identifying meanings with roles in analytic inferences is buying into the a/s distinction. So the cost of New Testament Semantics is buying into the a/s distinction. But, these days, practically everybody thinks that the a/s distinction is unprincipled. Indeed, it's widely thought that the discovery that the a/s distinction is unprincipled is one of the two most important achievements of modern philosophy of language; the other being precisely the theory that meaning supervenes on inferential role. If, then, we continue to assume that compositionality is non–negotiable, it follows that one of these foundational principles of the philosophy of language is going to have to go. Which one? And with what are we to replace it?

---

[4] There are causal theories of meaning around; see, for example, Skinner [1957], Dretske [1981], and Fodor [1991], among many others. And, like any account of meaning, each of them imply a corresponding notion of analyticity. But all these theories are externalist and atomist and thus offer no comfort either to CRT or to holism. They don't legitimate a construal of meaning in terms of analytic inference because they reject CRT; they don't reconstruct meaning in terms of inference at all.

3 What now?

This is where we ought to tell you how to go about mending the cracked foundations; we would if we could. Since we can't, however, we'll restrict ourselves to a few more or less jaundiced remarks on what we take the visible options to be and on what we think that adopting one or other of them is likely to cost.

The first possibility one might consider is to try resuscitating the a/s distinction. There is a certain face plausibility to this suggestion. We've been seeing that compositionality and analyticity come to much the same thing so long as you accept the New Testament view that meaning supervenes on inferential role. But, in fact, it's plausible that compositionality entails analyticity *whether or not* you accept New Testament Semantics. So, for example, it's hard to see how anybody could claim that the meaning of 'brown cow' is *compositional* while denying that the inference from 'brown cow' to 'brown' is *analytically valid.* If it is undeniable the meaning of 'brown cow' is constructed from the meaning of 'brown' and 'cow', it seems equally undeniable, and for the same reasons, that the inference from 'brown cow' to 'brown' is guaranteed by the linguistic principles that effect this construction. But an inference that is guaranteed by linguistic principles *just is* an analytic inference. In short, the very structural relations among the constituents of a sentence that ground its compositionality would appear to engender the analyticity of some of the inferences in which its constituents are involved.

So, to repeat, it appears that compositionality underwrites certain analyticities all by itself, without further appeal to the principle that meaning supervenes on inferential roles. A New Testament Semanticist might thus reasonably argue that if compositionality is non–negotiable, so too is analyticity; hence that the rejection of Quine's anti–analyticity arguments is independently motivated.[5] But if the revival of the a/s distinction can be defended, it looks as though ev-

[5] It's worth mentioning in this respect, that Quine says little or nothing about examples like 'brown cow → brown', where the analyticity of

erything is all right. We've seen that you can't have New Testament Semantics and compositionality *and the rejection of the a/s*, but you certainly can have the first two if you drop the third.

We emphasize, however, that the kind of a/s distinction that compositionality underwrites holds only between expressions and their *syntactic constituents.* It thus serves to distinguish, say, 'brown cow → brown' from 'brown cow → dangerous'. But it does *not* underwrite a distinction between, say, 'brown cow → animal', and 'brown cow → dangerous'. That is, it doesn't underwrite an a/s distinction among inferences that turn on the lexical inventory of the premises as opposed to their linguistic structure. It is, however, precisely the lexically governed inferences that make trouble for New Testament Semantics.

If the meaning of 'brown cow' derives from the meanings of 'brown' and 'cow' (as it must if the meaning of 'brown cow' is compositional) and if the meaning of 'cow' is its inferential role (as it must be if New Testament Semantics is right), then it must somehow be arranged that 'brown cow' inherits the inference 'animal' *but not the inference 'dangerous'* from the meanings of its constituents. But that requires that we must exclude from the semantic representation of 'cow' such information as, for example, 'cow → x kind of thing such that the brown ones are dangerous' (and analogously, we must exclude from the semantic representation of 'brown' such information as 'brown → x kind of thing such the cow ones are dangerous').[6] But to say that we must treat these inferences as excluded from the semantic representations of

---

an inference depends on the *structure* of its premises rather than their lexical content.

[6] Jim Higginbotham has pointed out to us that excluding dangerous from the semantic representations of 'brown' and 'cow' isn't sufficient to insure that 'brown cows are dangerous' comes out synthetic. This is because compositionality requires only that there be a function that determines the meaning of 'brown cow' given the syntax and lexical constituency of the expression. It doesn't, strictly speaking, also require that the semantic representation of 'brown cow' include the semantic representations of 'brown' and 'cow' (though many treatments of analyticity in the Kantian tradition have made this further assumption). This is alright with us; our point is that if 'brown cow' inherits dangerous

'cow' and 'brown' is just to say that they are both contingent; that they are *not* constitutive of the meanings of 'cow' and 'brown'. But to say that is to presuppose an a/s distinction that applies to lexically governed inferences. It is precisely an a/s distinction for lexically governed inferences that the familiar Quinean polemics are widely supposed to jeopardize.

We've been putting a lot of weight on the difference between analyticities that are engendered by the compositional structure of an expression and those that are generated by the meanings of items in its vocabulary. This distinction is, of course, undermined if you assume that there is a level of representation at which lexical items are semantically decomposed: to posit such a level is, in effect, to hold that lexical meaning is itself compositional. So, if there is lexical decomposition, then we can identify the meaning of an expression with its role in those inferences that are determined by its compositional structure *including the compositional structure of its lexical constituents.* The effect is to assimilate inferences like 'brown cow → animal' to inferences like 'brown cow → brown' since, at the level of semantic representation, both involve relations between expressions and their constituents. And both are distinguished from 'brown cow → dangerous' since, presumably 'dangerous' isn't a constituent of the representations of either 'brown cow' at *any* linguistic level.

But, of course, to resolve the present worries by taking lexical decomposition for granted simply begs the question against Quine. If lexical meanings were compositional, then lexical items would be definable; and it's part of Quine's story that there is no principled notion of definition. If Quine's arguments show anything, they show that there is no way to reconstruct the intuition that 'brown cow → animal' is definitional and 'brown cow → dangerous' isn't.

We conclude that, although there may be reasons for resuscitating the a/s distinction, the non–negotiability of compositionality isn't one of them. Compositionality licenses a distinction between 'brown cow → brown' and 'brown

---

from its compositional structure (in whatever way), then 'brown cows are dangerous' comes out analytic. Which is an unacceptable result.

cow → cow' but not between 'brown cow → dangerous' and 'brown cow → animal'. So the original problem stands: If compositionality isn't negotiable, then either there is an a/s distinction for lexically governed inferences (contra Quine) or inferential roles aren't what meanings supervene on (contra New Testament Semantics).

Our second point is that the present situation is rife with ironies. We remarked above that once you say that the meaning of an expression supervenes on its inferential role, it's hard to stop short of saying that the meaning of an expression supervenes on its *whole* inferential role. That is, it's hard to stop short of relativizing the meaning of an expression to the whole language that contains it, with the consequences that expressions in different languages are semantically incommensurable. There is a sort of Linguistic Idealist who delights in these holistic implications of New Testament Semantics; you see more than a trace of this sensibility among the 'cognitive linguistics' crowd, among connectionists, and, of course, among philosophers like Rorty, Putnam, Kuhn and Derrida. It now appears, however, that inferential role semantics *doesn't* have any holistic implications after all, so relativism loses however the argument turns out.

Here, roughly, is how the argument from inferential role semantics to semantic holism is supposed to go:

Premise 1. New Testament Semantics: The meaning of an expression is (at least partially) constituted by its inferential relations.

Premise 2. No A/S Distinction: There is no principled distinction between those of its inferential relations that constitute the meaning of an expression and those that don't.

Conclusion. Semantic Holism: The meaning of an expression is constituted by *all* of its inferential relations, hence by all of its role in a language.

There are, we think, lots of reasons to disapprove of this way of defending semantic holism; not least that it depends, apparently ineliminably, upon a form of slippery slope argument, and these are notorious for leading from true premises to false conclusions. (That is, it depends on arguing from 'There's no principled difference between the F's that are G and the F's that aren't' to 'Either none of the F's is G or all of them are'.) For present purposes, however, we're prepared to ignore all that. The point we want to emphasize is that the argument can't be better than its second premise; that is, the argument from inferential role semantics to semantic holism explicitly *depends on* denying the a/s distinction.

So, the situation is this: if the a/s distinction is principled, then the second premise is false and the argument is unsound. But if the a/s distinction is *un*principled, then the first premise is false. This is because, as we've been seeing, if compositionality is assumed, then meanings can be identified with inferential roles only insofar as the inferences that constitute a role are analytic; but the argument that there is no a/s distinction is the argument that there are no analytic inferences. We assume the law of excluded middle —either there is an a/s distinction or there isn't. Either way, then, the argument from inferential role semantics to semantic holism has to be unsound. We suppose that this demonstration should darken the skies of West Coast semantic holists. So be it; they need the rain.

The foundational problem, to recapitulate, is that you can't do all of the following: endorse inferential role semantics AND endorse compositionality AND abandon the a/s distinction. If compositionality is non–negotiable, that leaves you with only two options: endorse the a/s distinction or reject the idea that meaning supervenes on inferential role. We can't tell you which of these it would be best to do, but we do want to insist that wriggling isn't likely to get you off the hook. In particular, lots of cognitive scientists have hoped to reconcile themselves to the Quinean arguments by opting for a graded, or a contextualized, or otherwise denatured notion of analyticity. The idea is that, although Quine may have shown that the notion of *identity* of meaning is in trouble, it's still open that a notion of *similarity* of meaning might be evolved. It's supposed that grading the a/s distinction would allow us to have *both* compositionality *and* the supervenience of meaning on inferential role. New Testament Semantics would thus be vindicated, albeit in a sort of soft-edged version.

Nobody has shown that this can't be done; nor does anybody have the slightest idea of how to do it. It seems to us that the arguments against a principled notion of identity of meaning work just as well against a principled notion of similarity of meaning. But we won't try to make that case here.[7] Suffice it for present purposes to remark that the connections between analyticity and compositionality that we've been examining makes the search for a graded notion of analyticity look unpromising. Compositionality is, after all, a principle that governs relations *among meanings* (it governs the relation between the meaning of a complex expression and the meanings of its constituents). So, if your semantic theory reconstructs meaning in terms of analytic inference and if you have only a graded notion of analyticity, *then you will have to live with a graded notion of compositionality as well.* But what would a graded notion of compositionality be like? And, in particular, how would such a notion do what compositionality is required to do; viz. account for systematicity, isomorphism and productivity?

Wouldn't a graded notion of compositionality entail, at best, that a finite acquaintance with a language is adequate to *sort of understand* expressions not previously encountered? Or that if a language is capable of expressing the proposition that *aRb*, then it is sort of capable of expressing the proposition that *bRa*? Or that if the sentence *S* expresses the proposition *P*, then the constituents of *S* sort of express the constituents of *P*? But is there any sense to be made of such claims as, for example, that ('John loves Mary' sort of expresses the proposition that John loves Mary) only if 'John' sort of refers to John? These are muddy waters and we do not recommend that you wade in them.

It's worth mentioning that combining the denial of a/s with the espousal of CRT is perhaps the received position in cognitive science. The idea is that, on the one hand, concepts (meanings and the like) are prototypes; they're something like bundles of probable or typical traits. But, on the other

[7] See Chapters 1 and 7 of Fodor and Lepore (forthcoming).

hand, on one of the traits belonging to such a bundle is defining, so the a/s distinction fails. We take it that this option is closed when one notices the connection between compositionality and analytcity. If meaning is compositional, then either meaning isn't inferential role, or it is role in analytic inference. It's precisely because the bundles of inferences that constitute prototypes aren't analytic (and hence aren't compositional) that meanings belong to the brown prototype of the cow prototype. Indeed, the prototypical brown cow can be dangerous even if the prototypical cow is a pussycat.) So, again; either the meanings of 'brown' and 'cow' are their roles in analytic inferences, or they aren't their roles in inferences at all. If, as we suspect, Quine is right about a/s, then only the second disjunct remains open.

Here's where we've gotten to: Compositionality is non–negotiable; inferential roles are compositional only if the a/s distinction is tenable for inferences that are governed by the lexical inventory. So, if Quine is right about the a/s distinction, the only remaining alternative is to give up on the New Testament idea that the meanings of lexical items is constituted by their inferential roles. Barring some proposal for an entirely new kind of semantics, this requires going back to the Old Testament view that what makes 'dog' mean *dog* is some sort of symbol–world connection; perhaps some sort of causal or informational or nomological connection between tokens of the expression and tokens of the animal. There are, as we remarked above, various accounts of this currently on offer and, who knows, maybe one of them can be made to work.

The spin–offs for epistemology would be of some interest since the plausible candidates for the semantically relevant symbol to world relations all look to be *atomistic.* If 'dog' means *dog* because dogs cause '*dog*' tokenings, then it looks like your language could have a word that means what 'dog' means in English even if your language didn't have any other words that mean what English words do. Prima facie, in fact, it looks as though your language could have a word that means what 'dog' means in English even if it doesn't have any other words at *all.* All sorts of holistic arguments against the possibility of translation (and, more generally, against the 'commensurability' of languages and ideologies) would then appear to be likely to collapse. More darks skies on the West Coast.

We're not, to repeat the point one last time, actually urging you to follow any course. On the contrary, a lot of linguists will have good reason for not wanting to do so. For example, if Old Testament Semantics is essentially right, what becomes of such linguistic studies as 'lexical semantics'? You can't study the semantic relations among lexical items unless there are semantic relations among lexical items; and it's the burden of Old Testament Semantics that there aren't. Old Testament Semantics has it that semantic relations hold between lexical items *and the world* and *only* between lexical items and the world.[8] Old Testament Semantics has, therefore, no truck with semantic fields, lexical decompositions, conceptual networks and the like. If, indeed, what you mean by 'the semantic level of linguistic description' is the level of description at which sentences that differ only in synonymous expressions are identically represented, the natural way to read the Old Testament is as denying that there a animalis a semantic level of linguistic description. So generative semantics goes, and interpretive semantics *also* goes, and the highest level of linguistic description is, as it might be, syntax or logical form; viz. a level where the surface inventory of nonlogico–syntactic vocabulary is preserved. The God of the Old Testament is an austere God.

As previously remarked, our present purpose isn't to sell you any such chilly theology; just to point out that there seems to be this crack in the foundations of the structure in which semanticists and philosophers of language have recently been cohabiting. And to urge that somebody do something about it not later than sometime before the roof falls in.

[8] More precisely, the Old Testament view holds that, if there are semantic relations among lexical items, they are derived not from relations between their roles in the language but from relations between their connections to the world; ceteris paribus, two expressions connected to the world in the same way will be synonyms. But it is, of course, precisely the inferential relations among lexical items, and not their symbol–world relations, that the semantic level of lexical description has always been supposed to represent.

2

PHILOSOPHICAL ISSUES, 3
Science and Knowledge, 1993

Holism, Hyper–analyticity and Hyper–compositionality

Ned Block

1 The Old Hat

Fodor and LePore claim to have a new argument against inferential role semantics, a view that they regard as gospel not only in linguistics and philosophy, but also throughout cognitive science. I regard inferential role semantics as an embattled minority position that deserves to be much more widely accepted than it is.[1]

[1] This paper is a reply to Fodor's and LePore's (1992b), "Why Meaning (Probably) Isn't Conceptual Role", which is a version of their critique of Block (1986), Chapter 5 of their (1992a): "Ned Block: Holism and Conceptual Role Semantics". Though I have both versions in mind, I am keying my responses mainly to the paper in this journal, (1992b). Page references should be understood as indexed to that work unless (1992a) is specifically cited.

Before I try to defend inferential role semantics against Fodor's and LePore's new argument, I want to make some points about the old one, which I shall call the Old Hat.[2]

To begin with, inferential role semantics is the doctrine that meaning is inferential role. Supposing that one knows what it means to speak of the inferential role of sentences,[3] what does it mean to talk of the inferential role of words? In the first instance, it is sentences, not words, that one infers from and to, and thus it is sentences that *have* inferential roles. Nonetheless, we can think of the inferential role of a word as represented by the set of inferential roles of sentences in which it appears. The notion of the inferential role of a word (and likewise for other sub–sentential constituents) is a made–up idea, but this is an obvious way to make it up.

Here is the Old Hat. Suppose that I infer from 'This is a rattling snake' to 'This is dangerous —' for short, I will speak of inferring from 'rattling snake' to 'dangerous'. Now if we include *all* of a sentence's inferential liasons —that is, all of the inferences in which a sentence participates— in its inferential role, then, carrying this decision through to words, the inference from 'rattling snake' to 'dangerous' will be part of the inferential roles of —and hence part of the meanings of— the words 'rattling', 'snake' and 'dangerous' (I will ignore 'this' and 'is'). If I infer from 'rattling snake' to 'dangerous', and you do not, then we do not share the same meanings of any of 'rattling', or 'snake', or 'dangerous'. If we don't share the meanings of some of these words, then the meaning of 'Rattling snakes are dangerous' that I accept isn't the same as the meaning of this sentence that you do not accept. But then how can people ever disagree? The same reasoning leads to the conclusion that I can't change my own mind, since if I accept a sentence, and then later reject it, the meaning of the sentence that I accept isn't the same as the meaning of the sentence that I later reject. Further, apart from the example of perfect Twins, people are certain to dif-

---

[2]See Fodor (1987), pp. 55-56, Block (1986), pp. 629.

[3]This is something of a fiction, a useful one I think, on which both Fodor's and LePore's critique and my response, are based

fer in their 'snake' beliefs or in beliefs that are inferentially related to those beliefs, or in ... Thus, the meaning or content that you assert with 'Rattling snakes are dangerous' is most unlikely to be the same as the meaning or content that I assert with the same sentence, so agreement is almost as problematic for the inferential role semanticist as disagreement.

This is the problem that leads so many to abhor holism, a doctrine whose incarnation in inferential role semantics is the claim —assumed above— that all the inferences in which an expression participates are included in its inferential role.

2 The Crack

As wonderful as the Old Hat is, it is not what Fodor and LePore come to tell us about. They regard the Old Hat as "external to semantics", having to do rather with epistemology and ontology. They are after a new problem, a *linguistic* problem, the alleged Crack in the foundation of semantics. The Crack involves an inconsistent triad: inferential role semantics, compositionality and the rejection of the analytic/synthetic distinction. Compositionality is non-negotiable, they say, and since no one wants to accept the analytic/synthetic distinction (or so they say), we inferential role semanticists are in deep trouble.

The argument goes like this: with no analytic/synthetic distinction to appeal to, how can the inferential role semanticist find a principled basis for distinguishing among inferential liasons, including some in his inferential roles, while excluding others? He cannot postulate zero inferential liasons for very many words, since that would give all those words the same "zero" meaning. So, with no way to justify stopping between none and all, he must include all inferential liasons in inferential roles. But then inferential roles are not compositional. For if we infer from 'rattling snake' to 'dangerous', then this inference is part of the inferential role of 'rattling snake.' This inferential role, being part of the meaning of 'rattling snake', is required to be a function of the inferential roles of 'rattling' and 'snake', by the principle of compositionality. But, according to Fodor and LePore, the meanings of 'rattling' and 'snake' *in fact don't determine* that rattling snakes are dangerous. So if all inferential liasons are included in inferential roles, compositionality cannot be satisfied.

The only way out for the inferential role semanticist, say Fodor and LePore, is to give up holism and accept the analytic/synthetic distinction, for then we can nominate the analytic inferences as the ones that are *in* inferential roles, the synthetic ones being out. Analytic inferences, they say, are compositional. The upshot: the inferential role semanticist is caught between giving up compositionality on the one hand, and having the analytic/synthetic distinction forced upon him on the other.[4]

3 The Crack Depends on The Old Hat

I will offer a clarification, and then an objection. First, the clarification: Fodor and LePore seem to assume, along with many others, that Quinean objections to analyticity aside, the inferential role theorist has the option of appealing to analyticity as a way of discriminating the inferential liasons that are in inferential roles from those that are out. But if we stick to traditional ideas about the extension of 'analytic', there aren't *enough* analyticities.

Consider the putative analytic truths involving 'cat' —'Cats are animals,' 'Cats are living beings', 'Cats are grown up kittens', etc. The problem is that abstracting from the *words* 'cat' 'kitten', etc., appearing in these sentences, there is nothing here to distinguish 'cat' from 'dog'. Corresponding to 'Cats are grown up kittens', we have 'Dogs are grown up puppies'. Sure, 'Nothing is both a cat and a dog' can be used, but so can 'Nothing is both a dog and a cat.' Even if 'Cats are feline,' and 'Dogs are canine' are analytic, this

[4]The argument is neatly summed up in the last paragraph of their Part II.

is of no help without other analytic truths that distinguish 'feline' and 'canine'. Jerry Katz (1972) puts forward a view according to which words like 'feline' and 'canine' could be used to distinguish between the meanings of 'cat' and 'dog', but 'feline' and 'canine' would have to be primitives. That may be fine for his purposes, but an inferential role semanticist can have no truck with primitives, since he must take the meanings of *all* words to be given by their inferential roles.[5]

I've heard the response that the inferential role semanticist should see the difference in meaning between 'cat' and 'dog' in terms of a difference between the typical looks of the two animals, in what it is like to see them, or, more behavioristically, in the capacity to recognize them. But it would be hard to justify including the look of a dog in the inferential role of 'dog' while rejecting the inference from 'dog' to 'barks' because it isn't analytic.[6]

Conclusion: The inferential role semanticist has nothing to gain from appealing to (traditional) analyticity as a principle of distinguishing among inferential liasons.

It should be noted that the point just made and many other points in this paper involve examples of natural kind terms, where traditional claims of analyticity are widely thought to be relatively weak. Here I follow Fodor's and LePore's lead. Perhaps they have chosen the domain in part for its unfavorability to analyticity —since they wish to use Quine on analyticity as a club with which to beat the inferential role semanticist. Whatever their reason, a theory of what meaning is ought to be able to handle natural kind terms.

---

[5]This is all oversimple. First, inferential role semantics might be said to make every word a primitive. Second, one can imagine an inferential role semantics that appeals to different classes of inferences, giving a privileged status to inferences that correspond to traditional definitions.

[6]What should the inferential role semanticist do about the paucity of analytic inferences? I will argue later that inferential role semantics should be thought of in terms of narrow meaning and narrow content, and that this option can be used to avoid the problem. But even without this move, there are two obvious options: one is to include all inferential liasons in the inferential roles, and the other is to somehow extend inferential roles outside the head into the world.

Now I will turn to the main point against the Crack. Suppose that we include in the inferential role of 'snake' inferences such as that from from 'snake' to 'dangerous if rattling' and from 'rattling snake' to 'dangerous'. If we could include such inferential liasons, then the inferential role of 'rattling snake' would be a function of the inferential role of 'snake' (and hence a function of the inferential role of 'snake' and 'rattling' together), and the argument for the Crack would collapse, because compositionality would be satisfied. Remember that the Fodor–LePore argument depended on the claim that if we include all the inferences in which a sentence participates in its inferential role (i.e. we accept holism), we get non–compositional inferential roles, so if Fodor and LePore can't prevent including the inference from 'snake' to 'dangerous if rattling' in 'snake's inferential role, their Crack will crumble.

It is important to realize that no one is considering any alternative to including in the inferential role of a word all the inferences involved in inferential roles of sentences which contain that word. Fodor and LePore don't mention any alternative, nor as far as I know is any alternative ever discussed in the literature. As I said at the outset, it is sentences that in the first instance *have* inferential roles, and the natural way of thinking of the inferential roles of words and other sub–sentential constituents of sentences is as constructions out of the inferential roles of sentences. Whatever inferences one decides to include in the inferential roles of sentences will be automatically carried through to the level of words.

Fodor and Lepore argue that meanings are compositional; but inferential roles are not compostional; so meanings cannot be inferential roles. But the idea that "compositionality is an embarrassment" (p. 23) for inferential role semantics is most peculiar. A genuine issue about compositionality arises, for example, with respect to externalist semantic theories, where we have an independent purchase on the reference or satisfaction conditions of words, and the truth conditions of sentences. There is a real issue of whether the latter is a function of the former. But in the case of inferential role, where we have no independent purchase on the inferential role of words, no such issue arises. In the context of inferential role semantics, talk of accepting or rejecting compositionality is hard to understand without some non–compositional proposal and some reason why any inferential role semanticist might accept it. The most obvious non–compositional proposal would be to have *two* Official Lists (or Principles for Constructing Lists) of inferences to be included in inferential roles, one for sentences, and another for words (or more Lists for other subsentential constituents). I can't think of a reason why an inferential role semanticist would want to do that, and Fodor and LePore don't even hint at anything to do with this issue. Indeed, Fodor and LePore make it appear that there is a "compositionality problem" for inferential role semantics only by explicitly assuming that the meaning of 'brown' is the property of being brown, and the meaning of 'cow' is the property of being a cow (p. 23), which is a *non*–inferential–role account.

Clarity about compositionality requires clarity about the distinction between compositionality and holism. Holism, for inferential role semantics, is a matter of *how many* inferences are included in the Official List of inferential roles. If all inferences are included, we get holism. The example of including in the Official List the inference from 'This is a rattling snake' to 'This is dangerous' is meant to be a proxy for holism. Compositionality, by contrast, has to do with our stipulation of what the inferential roles of words are to be, *given* that one has already decided on the Official List of inferential roles of sentences.

In sum, Fodor and LePore are right that there is an issue of whether to include the inference from 'rattling snake' to 'dangerous' in our Official List of inferential roles. But if we *do* decide to include it in the Official List for sentences, *of course* we should include it in the Official List for words. So the real issue is one of holism, not compositionality.

Before I get to their argument, let me make some further points about compositionality. If holism is true, the semantic identity of each word in the language carries information concerning the semantic properties of all the words in the language. Holistic compositionality is hyper–compositionality. It would be natural to respond by framing a definition of 'compositionality' that would include the non–holistic compositionality that Fodor and LePore have in mind, while excluding hyper–compositionality, in which the meanings of each of 'snake', 'dangerous' and 'rattling' contain the information that rattling snakes are dangerous. But though this would not be hard, it would be pointless, since hyper–compositionality is genuine compositionality, though an extreme form of it. To redefine 'compositionality' to exclude hyper–compositionality, and then complain that inferential role theories are not compositional would be an ad hoc verbal maneuver of no interest.

We can clarify hyper–compositionality further by giving a more explicit proposal about the semantic values of words. First, let the semantic value of a sentence be an ordered pair whose first member is the set of inferences *to* that sentence and whose second member is the set of inferences *from* that sentence. (On many views of what is to be included in inferential roles, these will be infinite sets.) Then the "semantic value" (in the sense of the semantic property the theory assigns) of a word could be taken to be the set of semantic values of sentences containing that word, i.e. a set of ordered pairs, one for each sentence containing the word.

These semantic values are compositional in the sense used here, *semantic* compositionality, namely there is a function from the semantic values of parts to the semantic values of wholes. (This is the sense that is used in formal semantics, and that appears in many places in Fodor and Lepore (1992a) —see the Glossary, for example.)[7]

There is another sense of 'compositionality', *psychosemantic* compositionality, in which hyper–compositionality clearly does not involve compositionality. For a theory to be compositional in this sense is for it to tell us how a person *represents* the meanings of words and uses them to form representations of the meanings of sentences. A holistic IRS does not do this

[7]The function will also be computable: If one wants to "compute" the semantic value of 'Water floats ships' from the semantic value of 'water', one can search the ordered pairs in the semantic value for 'water' until one finds one in which 'Water floats ships' is repeated (in one of the sets in the pair): That ordered pair is the semantic value of the sentence.

by itself (nor do other non–psychosemantic theories, possible worlds semantics theories, for example), so it requires supplementation by a psycho–semantic theory. It is psychosemantic complexity that is needed to explain such psychological phenomena as systematicity and productivity. As Fodor (1987) points out, it seems to be a fact about humans that if they can frame the thought that Mary loves John, then they can frame the thought that John loves Mary. A theory of how people represent the semantic and non–semantic properties of the "parts" of these thoughts has the potential to explain such facts.

To the extent that a hyper–compositional theory seems wildly counter–intuitive, this in my experience is due to a confusion of these two senses of 'compositional'.

Hyper–compositionality is dramatically illustrated by attention to the probabilistic inferential role theory that Fodor and LePore mention. According to this theory, the inferential role of a sentence is a matter of its conditional probability relative to *every* other sentence.[8] It is this conditionalizing on every other sentence that brings in the holism. Fodor and LePore argue against the compositionality of these probabilistic inferential roles as follows. They imagine that the (subjective) probability that a rattling snake is dangerous is high, whereas the probability that a snake is dangerous is low (most snakes are harmless), as is the probability that a rattling thing is dangerous. (Actually, they put this in terms of their brown cow example.) They argue that since the first is high, whereas the second and third are low, the semantics of the first can't be a function of the semantics of the second and third. There is some danger in thinking of compositionality in terms of relations among sentences —as opposed to the relations between sentences and subsentential constituents. But going along with this fiction, lets suppose that compositionality requires that the semantics of (1) be a function of the semantics of (2) and (3).

---

[8] I should mention that Hartry Field, in the article that Fodor and LePore mention says he is giving a notion of sameness of meaning, not meaning itself. Also, since these are subjective probabilities, the criterion of sameness of meaning is intrapersonal, not interpersonal.

1. This rattling snake is dangerous.

2. rattling thing is dangerous.

3. This snake is dangerous.

Now Fodor and LePore see the issue as settled by the fact that (1) has a high probability, whereas (2) and (3) have low probabilities. Without pausing to consider why a compositionality function can't take two low probabilities into a high probability, we can see an important problem. The probabilistic semantics for these sentences is a matter of their *conditional* probability relative to every other sentence. This semantic entity is *a function from sentences to probabilities, not a simple probability.* This function in the case of (2) *includes* (2)'s *high* probability conditional on 'This is a snake'. Likewise, this semantic function in the case of (3) includes its high probability relative to 'This is rattling.' In short, the probabilistic meaning function for (2) and (3) both *include* the information that rattling snakes are dangerous.

Leaving the topic of probabilistic inferential role, let us return to the main question of this section: just what do Fodor and LePore have to say against including the inference from 'rattling snake' to 'dangerous' in the inferential role for 'snake'? Recall that without a way of precluding this option, the argument that including all inferential liasons in inferential roles yields non–compositional inferential roles will fail, and they will not be able to conclude that the inferential role semanticist has no choice but to accept analyticity. So their whole argument depends on this step.

No argument is given in (1992b), but there is one in (1992a): "...there is a way to reconcile analyticity with holism —namely by accepting a semantics that represents *every* inference as analytic...the idea that all inferences are compositional is the idea that all inferences are analytic, and the idea that all inferences are analytic, though it, too, is compatible with holism, is, on independent grounds, perfectly mad". (181-182) So the claim is that if we include the inference from 'rattling snake' to 'dangerous' in the inferential role for 'snake', we make 'Rattling snakes are dangerous' analytic, (or the inference from 'rattling snake' to 'dangerous' analytic) and that is bad.

Let us postpone the issue of *why* including the inference is supposed to make 'Rattling snakes are dangerous' analytic, and ask first why, *if* this sentence is analytic, that is supposed to be bad?

The sum total of the argument in (1992a) is that this idea is "preposterous on the face of it" (164), "patently preposterous" (174), that it is not "possible to take seriously" (174), that "surely this is preposterous" (182), that it is "perfectly mad" (182), and "incapable of being taken seriously" (183), and that it is "...an option which Quine, quite sensibly, didn't even bother to consider..." (183). No other argument appears in either publication.[9] I have three points to make about all this. First, this sort of confidence about what is and is not analytic does not fit with their attempt to beat the inferential role semanticist about the head and shoulders with the club of Quinean rejection of analyticity. I will come back to this point. Second, is all the name-calling in part a way of invoking the Old Hat? If 'Rattling snakes are dangerous' is analytic in my idiolect but not in yours, then, they might say, how could it be that my accepting this sentence and your rejecting it amount to a disagreement? Adding analyticity to the Old Hat in the "no analyticity" version presented at the outset of this paper perhaps makes it more dramatic, but I don't see that it adds any substance. Finally, philosophers are wont to take their intuitions about concepts very seriously. And there is some justification for this. Our concepts of consciousness, agency, belief and so on play an important role in our lives, and we are interested in an elucidation of them, warts and all. But analyticity is a technical concept whose warts are of only technical interest. In some arenas, what we want by way of theory is a "good fit" to intuitions, but this methodology is not always appropriate. In semantics, what we want is a theory with some power and interest. If we have to get rid of intuitions about 'analyticity', etc.,

[9] Well, actually there is an argument that has the role of an aside in (1992a), p. 182 to the effect that this sort of holism would preclude the Quine/Duhem thesis. I won't discuss this point further.

so be it. (I would say the same about 'meaning', but this is more controversial.) These intuitions are largely the product of the discipline itself.

Katz (forthcoming) argues that Kant and Locke thought of analyticity as a narrow meaning containment relation. Frege introduced the current use in which truth plays such an important role. Thus our intuitions concerning 'analyticity' may vary with prevailing theory.

How might Fodor and LePore back up the claim that it is preposterous to regard the inference from 'rattling snake' to 'dangerous' as analytic? One possible Fodor–LePore answer would be that it is preposterous because if we were to regard all inferences as analytic we would not be able to recognize the existence of disagreement, change of mind, etc. If they were to give this answer, the Crack would be little more than the Old Hat.

What other answer could they give? Others might say that 'Rattling snakes are dangerous' isn't analytic or that there is no matter of fact as to whether it is analytic because nothing is analytic or because there are no facts of analyticity, or because analyticity is a hopeless confusion.[10] Interestingly, Fodor and LePore have closed off these avenues of response. They endorse the analyticity of "quasi–syntactic" inferences like that from 'brown cow' to 'brown', or to 'cow'. (I call these 'quasi–syntactic' because they depend on the repetition of a word.) The Quinean position that has been so influential rejects the very notion of analyticity as a hopeless confusion.[11] By accepting *some* analyticities, Fodor and LePore pull the Quinean teeth from their critique.

---

[10] Boghossian (forthcoming) distinguishes among a number of such anti–analyticity theses.

[11] Regardless of how Quine himself should be interpreted, I think that this is the doctrine that he has been taken to have made plausible. Horwich (1992, forthcoming) maintains that Quine himself does not hold the view that I am labeling 'Quinean', viz., the rejection of analyticity as a hopeless confusion. Horwich argues that in Quine's use of the "reducibility to logical truth via definitions" notion of analyticity, he is committed to the analyticity of logical truth. Fodor and LePore (1992a),p. 245 express a similar view.

What I am leading up to is this: Given that they accept *some* analyticities, how can they reject the analyticity of 'Rattling snakes are dangerous' without saying something like "Well, *in fact* 'dangerous' just isn't part of the meaning of 'rattling snake'"? Since the issue of whether 'dangerous' is or is not part of the meaning of 'rattling snake' is part of the very issue at hand (because we are discussing their argument against holism, i.e. against including all inferential liasons in inferential roles), this plainly would be begging the question. They are up to their necks in analyticity, so the issue about whether all inferences are to be included in inferential roles is just the issue of holism itself. The cash value of their argument is just the bare claim that holism itself is preposterous, and this claim has to be argued for, not just asserted. This is half of my response to their objection.

Well, what if Fodor and LePore were to try to evade the objection just given by becoming Quineans, rejecting analyticity as a confusion? This, they cannot do. Recall that I have been postponing the issue of why including the inference from 'rattling snake' to 'dangerous' in the inferential role for 'snake' is supposed to make 'Rattling snakes are dangerous' analytic, concentrating instead on why, if this sentence is analytic, that would be bad. But of course the two questions are closely related. Were Fodor and LePore to give a Quinean response to the latter question, their approach to the first would dissolve. That is, rejecting analyticity as a confusion would preclude their claiming that the holist has to regard 'Rattling snakes are dangerous' as analytic. This is obvious enough, but the points to be made in the next section will articulate it.

Further, their views on compositionality and its relation to analyticity would make it difficult for them to reject analyticity. "...it's plausible that compositionality entails analyticity *whether or not* you accept New Testament Semantics...In short, the very structural relations among the constituents of a sentence that ground its compositionality would appear to engender the analyticity of some of the inferences in which its constituents are involved". (p. 338)

I've alluded to the fact that Fodor and LePore tend to equivocate on 'analyticity'. Their official notion is truth in virtue of meaning.[12] But sometimes they seem to have in mind a distinct sense of 'analyticity', namely what one might call "traditional analyticity", in which the extension of the term is tied to traditional examples, especially lexical cases like 'Cats are animals'. The point I've been making involves the official notion of analyticity. The holist whom they are arguing against says that the inference from 'rattling snake' to 'dangerous' is to be included in the meaning of 'snake'. To deny this on the ground that that the inference is not in fact analytic is just to beg the question, if analyticity = truth in virtue of meaning. But what if the extension of 'analyticity' is pegged to traditional examples? This would block their argument in another way, for their argument starts with the claim that to include the inference from 'rattling snake' to 'dangerous' in the meaning of 'snake' would be to make this inference (and the sentence 'Rattling snakes are dangerous') *analytic.* If analyticity is restricted to traditional examples, this claim would certainly be false.

In sum, the Crack rests on the argument that if we include the inference from 'rattling snake' to 'dangerous' in the inferential role of 'snake', we make this inference (and 'Rattling snakes are dangerous') analytic, and that is bad. (Of course, this example is just a proxy for talk of including *all* inferences in inferential roles.) Here I have dealt with the latter, the *bad* part; the next section is about the former, the *analyticity* part.

What was my response to the *bad* part? I said that their rejection of the analyticity of 'Rattling snakes are dangerous' without argument amounts to the rejection of holism, without argument. But do *I* think that 'Rattling snakes are dangerous' is analytic or not? My answer, given below, is that inferential role semantics of the sort I favor is committed to the narrow meaning of 'dangerous' being included in the narrow meaning of 'rattling snake', so 'Rattling snakes are dangerous' has the narrow analog of analyticity. This in-

[12] "... the analytic/synthetic distinction is unprincipled. This means ... that there aren't any expressions that are true or false solely in virtue of what they mean" (p. 20).

volves no commitment to the analyticity of 'Rattling snakes are dangerous', as I will show below.

4 The Plausible–Sounding Principle, and Compositionality

Let us now turn to Fodor's and LePore's argument that including the inference from 'rattling snake' to 'dangerous' in the inferential role of 'snake' makes this inference analytic. It is not too difficult to see what the argument is supposed to be. Fodor and LePore say: "For an inference to be analytic is for it to be warranted by the meanings of its constituents. But, according to New Testament Semantics, meanings are inferential roles. So for an inference to be analytic is for its warrant to be determined by the inferential roles of its constituents". So if the inference from 'rattling snake' to 'dangerous' is part of the inferential role and therefor the meaning of 'snake' (as the holist inferential semanticist claims), then this is an analytic inference. (This reasoning appears in the voice of an interlocutor on p. 26, but reappears in the authors' voice on p. 27.)

The argument presupposes something like the following principle:

The Plausible–Sounding Principle. Inferences that are part of inferential roles must be regarded by the inferential role theorist as analytic. For these inferences are what are taken to constitute meaning, and inferences that constitute meaning are analytic.[13]

Fodor and LePore see the Plausible–Sounding Principle as a consequence of their view that, for inferences, "composi-

[13] Note, incidentally, that the Plausible–Sounding Principle cannot use 'analytic' as restricted to traditional sorts of examples, for then it would be of no use in arguing that the holist inferential role semanticist is committed to the analyticity of 'Rattling snakes are dangerous', since this isn't of the type of the traditional examples.

tionality entails analyticity" (p. 26).[14] The idea is that "...if "brown cow" → "dangerous" is *compositional* [i.e., if this inference is part of the inferential role of 'brown' or 'cow'], it follows that it is also *analytic*...the cost of representing an inference as compositional is that you then have to represent it as analytic". (1992a, p. 181-182, italics theirs). So if an inference is included in the inferential role of a sentence (such as the inference from 'This is a rattling snake' to 'This is dangerous'), and therefore is included in the inferential role of the constituent words (e.g. is part of the inferential role of 'snake'), then this inference is analytic. Hence the Plausible–Sounding Principle.

The rest of this paper is about the Plausible–Sounding Principle. Though Fodor and LePore don't make much of it, I see it as the corner–stone of the general line of thought that they are running to the effect that inferential role semantics runs into trouble with analyticity. Should they be tempted to pull back from a commitment to it, they would have to show how, without it, they can derive the analyticity of the inference from 'rattling snakes' to 'dangerous' or of 'Rattling snakes are dangerous' from the claim (of holistic inferential role semantics) that the inference from 'rattling snake' to 'dangerous' is to be included in the inferential role of 'snake'.

Before I get into a critical examination of the Plausible–Sounding Principle, I have a few remarks about analyticity. You will recall that Fodor's and LePore's official notion of analyticity is truth in virtue of meaning. As I mentioned, they also use 'analyticity' so as to restrict it to traditional examples. Now if it is analyticity in the latter sense which they are trying to foist on the inferential role semanticist, they *never* had an argument that the inferential role semanticist is committed to analyticity in the first place.

To see this, suppose that Fodor and LePore are right that it is unacceptable to regard 'dangerous if rattling' as part of the meaning of 'snake', and thus that the inferential role

[14] See fuller quotation five paragraphs back. They go further, saying "...for inferences, analyticity and compositionality are the same thing..." (1992a, p. 185, italics theirs).

semanticist is unable to use "All of them" as a principle of inclusion of inferential liasons in inferential roles. (This is the step in the argument that I have been worrying about up until now.) They go on to claim that once holism is rejected, the inferential role semanticist has nothing left to fall back on as a principle of inclusion except analyticity. Now if analyticity here means 'true in virtue of meaning', or even 'reducible to logical truth via substitution of definitions', the argument has a certain prima facie appeal (though I will be arguing that there is an unnoticed assumption here that should be rejected). The idea would be that all and only inferences that the inferential role semanticist can independently justify as analytic should be included. But if 'analytic' involves a commitment to the *extension* of 'analytic' as traditionally viewed (including, for example, to 'Cats are animals'), then the idea that the inferential role semanticist has *nothing else to fall back on* seems more than strained. It is hard to fault the claim that the inferential role semanticist should have a *principle* of distinguishing between inferential liasons that are in from those that are out. But it is another *more doubtful* thing to say that the only such principle is analyticity. (I will be arguing that analyticity won't do.) And it is still another *much more doubtful* thing to say further that the only option is to accept traditional ideas of what is included and what is excluded from that category, especially since as I mentioned earlier, there aren't enough of these traditional analyticities. Any holist who accepts the criterion of analyticity is committed to regarding 'Rattling snakes are dangerous' as analytic, and thus this holist is committed to rejecting the traditional extension of 'analytic'.

In sum, if Fodor and LePore stick to their official sense of 'analytic', then their emphasis on compositionality is misplaced, for the Plausible–Sounding Principle that they have to buy into anyway would, if true, require the inferential role semanticist to accept the analyticity of whatever inferences he includes in inferential roles. But if they use 'analytic' so as to involve a traditional idea of the extension of 'analytic', then their argument that the inferential role semanticist is committed to analyticity loses its plausibility.

Returning to the Plausible–Sounding Principle, I should add that the Fodor–LePore argument based on it is a special case of a more general argument developed in Boghossian (forthcoming), which I object to on the same ground as I object to the Plausible–Sounding Principle.[15] Boghossian notes that philosophers have accepted Quine's argument against the analytic/synthetic distinction, but not Quine's attack on meaning itself. He argues that anyone who accepts determinate facts about what means what is committed to analyticity, so one has either to embrace both of the Quinian doctrines or neither of them.

The Plausible–Sounding Principle holds that for the special case of inferential role semantics, determinate meaning facts engender analyticity, but the considerations that favor the Plausible–Sounding Principle are general, and would apply to other theories as well (see the 2nd paragraph of part III for partial acknowledgement of this point). But if Fodor and LePore accept the general form of this argument, they are in trouble, for they themselves accept determinate meaning facts as part of their Old Testament semantics.[16]

I will give an argument against the Plausible Sounding Principle. It will show how an inferential role semanticist can have holism plus compositionality without analyticity, and without succumbing to the Old Hat. It will also apply against Boghossian's argument. My point is very simple. It is that inferential roles theories are much more plausibly taken as theories of narrow meaning than of wide meaning, and —this is the punch line— determinate meaning facts about *narrow* meaning do not engender analyticities. So the Plausible Sounding Principle (that inferences that constitute meaning are analytic) is wrong —for narrow meaning. This point also blocks the Old Hat, the argument discussed at the outset that holistic inferential role semantics cannot acknowledge agreement and disagreement.

---

[15] I accept Boghossian's claim for *wide* meaning, but not for narrow meaning. See below. Some of the points here made against Fodor and LePore were arrived at independently by Boghossian.

[16] Bob Stalnaker made essentially the same point in comments sent to Fodor and LePore.

It may be said that Fodor and LePore needn't be seen as arguing against narrow inferential role theories. Perhaps they are best taken as refuting a wide inferential role semantics in which the roles are seen as extending into the world? (This might be said, but actually their (1992a) is explicit about being directed against both narrow and wide inferential role semantics. Indeed, they are at pains to point out that their argument doesn't depend on assuming that inferential role semantics is only an account of narrow content. See p. 168) So how can I defend wide inferential role semantics against their attack by defending narrow inferential role semantics? It suffices to say that if their attack can be shown not to apply to the most plausible version of inferential role semantics, why should those who favor inferential role semantics be disturbed by it? The ball is (or soon will be) in their court on this issue.[17]

I should mention at this point that I am not a True Believer in inferential role semantics or in narrow meaning or content. Rather, I think both are interesting and promising theoretical options that will stand or fail together, and that no one has succeeded in working out successfully. Though I am not a

[17] See Block (1987) for a discussion of the distinction between narrow and wide inferential role semantics. I discuss in that paper a "two factor" inferential role semantics in which one factor is "short–arm" inferential role that stops at the skin. This short–arm inferential role is meant to capture narrow meaning or content. The other factor is a referential/truth–conditional factor. I argue that this sort of theory is equivalent to a "one–factor" theory of the sort favored by Harman (1982), in which the one "long–arm" factor extends into the world. The holism issue reveals a difference between two different inferential role theories of wide content, though the difference doesn't line up exactly with the distinction between one factor and two factor theories. The two factor theory is equivalent to the one factor theory only if they are non–holistic, so my (1987) was mistaken. On the two factor theory, the holist has the option of making only the narrow factor holistic —this is the view defended here, a view that avoids the Old Hat. The one factor theorist, by contrast, if he is also a holist, will most naturally treat his entire factor holistically, yielding a distinct theory that succumbs to the Old Hat. In sum, the two most natural holistic inferential role theories of wide content are (a) a two factor theory that treats only the short–arm roles holistically, and (b) a one factor theory that treats the long–arm roles holistically.

True Believer, I do think that inferential role semantics is not refuted by the arguments given by Fodor and LePore, and that is the position I see myself as defending here.

There are many differences between thought and talk, and between the content of thought and the meaning of talk. But in what follows, I will tend to ignore these differences.

5 Two Factors and Disagreement

Let's agree that people do disagree, change their minds, etc. Now there is one obvious version of inferential role semantics that allows us to have our cake and eat it too —have holism and also disagreement, changes of mind, etc.

The version I have in mind involves a "two factor" theory, one that combines the Old and New Testament approaches.[18] According to one way of thinking of the two factors, there is an internal content factor —narrow content— that is holistic. And there is an external content factor, the theory of which is to be filled in, perhaps, by Fodor's own externalist account. Some object to this sort of view because they have no stomach for narrow content. This is not the place for an attempt to defend narrow content. I will rest with an ad hominem argument, namely, how can Fodor, Mr. Narrow Content, himself, say this?

Here's how the two factor view avoids the Old Hat: In thinking about disagreement, agreement, change of mind, and such, we let the wide contents individuate the propositional attitude contents. Suppose, temporarily, that we think of wide contents as purely referential/truth-conditional. If I first assert, then, in an about-face, deny 'Toads cause warts', we can individuate the contents as the one factor externalist would do so, concluding that I deny what I earlier asserted,

---

[18]See Dretske, 1988, Ch. 6.4, especially Note 2 on p. 150, for an indication of why one who starts with a single factor externalist theory (of the Old Testament sort that Fodor has expounded and that Fodor and LePore seem to be pointing to) might find himself moving in the two factor direction. See Harman (1982), Horwich (1990) for one factor accounts. See also Devitt (1990).

despite the changes in narrow contents. In other words, for some purposes of attitude attribution, the equivalence relations among contents would ignore the narrow contents.

Similar points about individuation of computations are familiar. We can type–identify computations by function computed, ignoring the specific algorithm deployed and its implementation, or by the algorithm, or by the implementation. We can type identify bathtubs architecturally —in terms of size, shape, weight, and decorative properties. Alternatively, we can type identify them economically, in terms of initial price, usable life, and thermal insulation, ignoring architectural features. Equivalence relations can be based on any of these features.

Martin Davies has suggested that I point out that essentially this idea about individuation of content can be appreciated without any commitment to narrow content. Suppose that the way that you think about Hesperus differs from the way I think about Hesperus. Nonetheless, the word 'Hesperus' reasonably may be taken to have the same meaning in our mouths. If meanings were individuated as finely as thought contents, no two people would be likely to share meanings. So we individuate in a more coarse grained manner. But we will also want to maintain that 'Hesperus' differs in meaning from 'Phosophorus'. The point is that we can think of thought contents as determining reference and truth conditions, but still being much more fine grained than the truth conditions that they determine. And we can think of meanings as determining truth conditions while still being more fine grained than the truth conditions, but not as fine–grained as the thought contents. We have a many–one relation between thought contents and meanings, and a many–one relation between meanings and truth–conditions. Thus we can see two different equivalence relations on detailed thought contents. In thinking about agreement and disagreement using this apparatus, we would individuate in terms of meaning or in terms of truth conditions, depending on context, though some contexts might also elicit individuation in terms of fine–grained thought contents.

Adding in narrow content complicates this simple hierarchical structure, because narrow contents are both more and less coarse-grained than truth-conditional contents. My 'Hesperus'-thought has the same narrow content as my Twin's 'Hesperus'-thought, despite their truth conditions being different —if the star referred to as 'Hesperus' here is distinct from the one referred to on Twin Earth by the same name. Though narrow contents are in this respect coarser grained than truth-conditional content, narrow contents are also finer grained, since my 'Hesperus'-narrow content differs from both your 'Hesperus'-narrow content and my 'Phosphorus'-narrow content.

In the last four paragraphs, I supposed that wide contents are purely referential/truth-conditional. On this way of talking, the wide content of the belief that we express with 'Hesperus shines' would be the same as the wide content of the belief that we express with 'Phosphorus shines'. On this familiar view of wide content, wide content is entirely within the purview of the external factor of the two-factor theory. The cognitive difference between the beliefs expressed by the two sentences just mentioned could, on this view, be captured in terms of different cognitive relations to a single wide content. (See Fodor, 1990, for a view of this sort.) A distinct but also familiar terminology regards wide content as not entirely referential/truth-conditional. This view gives the beliefs expressed by 'Hesperus shines' and 'Phosphorus shines' different wide contents. On this view of wide content, narrow content contributes somewhat to the individuation of wide content. Specifically, the difference in the inferential roles of the words 'Hesperus' and 'Phosphorus' makes for a difference of wide content, despite the fact that the two words refer to the same thing. The point depends on the different roles of the two words, but of course a single word can have two different roles as well.[19] I don't want my reply to Fodor and LePore to depend on either view of how wide contents are to be individuated. I can exploit whatever principle of individuation is favored by those who dislike holism.

[19] Pierre visits London twice, once in a foul mood in drizzling rain, again, as if to a different place, in high spirits and brilliant sunshine; He assumes that there are two places in England called 'London', one sparkling, one dreary. See Kripke (1979), Loar (1987).

Of course, since narrow contents are holistic, whereas truth-conditional contents are not, narrow contents will be individuated much more finely (even though they will also be individuated more coarsely, as in the Twin Earth examples). So: many differences in narrow content will not make for differences in truth–conditional content. It is reasonable to ask what narrow content is *for* (and also what truth–conditional content is for). Answer: narrow content has a role in psychological explanation[20] For purposes of certain kinds of psychological explanation, narrow content differences matter despite the fact that they don't make for differences in truth–conditional content, as Loar (1987) points out. By contrast, truth–conditional content attribution is useful for communication and other contexts where information is important, and where psychological differences don't matter. If the word gets around that the announcer has said that Bigbucks won the Kentucky Derby, all concerned can pick up the information without attending to how the announcer conceives of these things. The announcer's narrow contents matter to predicting and explaining his behavior, but not to the information he has communicated. According to this picture, it is wide contents —or perhaps referential/truth–conditional contents, if these differ from wide contents— that are attributed with *that* clauses.[21] Narrow contents, by contrast, are not paired 1–1 with wide contents, and hence not "straightforwardly" attributible. We can, however, speak of my narrow content for "Water puts out fires", or yours. "You can't, in absolute strictness, express narrow content; but as we've seen, there are ways of sneaking up on it". (Fodor, 1987, p. 51)

I have mentioned three categories of content: narrow content, truth–conditional content and wide content. I also mentioned two distinct views of wide content. According to the

[20] Note that I am *not* claiming that narrow content is the only content that has a role in psychological explanation. Explaining behavior in terms of narrow contents does not preclude *also* giving some explanations of behavior in terms of wide contents.

[21] I differ from Loar in *accepting* wide contents; Loar confines himself to talk of wide attributions.

first, wide content reduces to truth–conditional content. On this view, Oedipus both accepts and rejects the wide content expressed by 'Oedipus is married to his mother', but via different modes of access to this wide content. According to the second view of wide content, Oedipus accepts one wide content (expressed by 'I am married to Jocasta'), but rejects a distinct content (expressed by 'I am married to Mom'). Narrow content has a home in psychological explanation, truth–conditional content has a home in communication. I do not take a view on the utility of wide content to the extent that it differs from truth–conditional content.

6 Analyticity, Finally

"Not so fast", Fodor and LePore might say. "What about all those analyticities? The analytic truths involving narrow contents that you are committed to *preclude the option* of individuating contents in terms of truth conditions, given your holistic narrow contents. You can't just pretend your narrow contents aren't there, since analyticity has consequences for truth. Suppose I used to think that toads cause warts, but now, having changed my mind, I accept that toads don't cause warts. The externalist says that the content of my earlier 'Toads cause warts' is the same as what I now deny with 'Toads don't cause warts'. But you *can't* say that, since your views about constitutivity of narrow contents commit you to the claim that both the original assertion and its later denial are analytically true. How could one content be such that its assertion is analytically true, and its denial is analytically true too? Analytic truths are true in virtue of meaning, and since the meanings of some of the words must have changed, so have the contents".

The Fodor and LePore reply that I am imagining is that the putative analytic truths involving narrow contents preclude treating the truth–conditional content of 'Toads cause warts' as the same when asserted as when denied. In short, I am imagining Fodor and LePore to be making a last–gasp appeal to a version of the Old Hat. In doing this, they will be conceding that the Crack does not stand on its own, but that will be cold comfort to me, if I am defeated by the Old Hat. If their objection is right, it will disable my refutation of the Plausible–Sounding Principle and my reply to Boghossian too.

Here's how I avoid this objection. I point out that narrow contents are *never* analytic. They aren't even the *sort* of thing that can be analytic. To see this, note that narrow contents are never *true*, and hence have no truth conditions. Analyticity requires truth —truth in all conditions, so something that can't be true can't be analytic either.[22]

I think it will help, but only as a first step in discussing this matter, to consider my utterance of 'Feynman was a genius'. It is true. But now consider my Twin on a Twin Earth where Twin Feynman is not a genius, but only a front man for a shy genius who feeds Feynman all his lines. This is done sufficiently smoothly that no causal signals from Twin–Feynman to *my* Twin have been any different from the causal signals from the real Feynman that have reached me. My Twin says 'Feynman was a genius', but what he says is false. My Twin is as much like me as you like. So we have the same narrow contents. Conclusion: thoughts and utterances with the same narrow content can be true in one environment, false in another. Narrow contents have no truth values, and hence no truth conditions.

This point is only a first step because it can reasonably be objected that all the Feynman story shows is that truth conditions are relative to context. Such narrow contents might still have an analyticity–like property, namely, yielding a truth (in virtue of meaning) *in all contexts.* Let us call

---

[22]Note that narrow content inferences cannot be regarded as inferences in any sense that involves *truth.* (See Block, 1987, for more on this point.) But this is OK with Fodor and LePore, since they say that they are happy to accept a causal interpretation of 'inference' —so long as it is not an interpretation of *analytic* inference. Note also that the proposal for defining the inferences that are in inferential roles as the analytic ones is ridiculous if inferential role semantics is taken as a theory of narrow content, for there can't be any analytic narrow content inferences.

this property *hyper–analyticity.*[23] A hyper–analytic narrow content makes any wide content of which it is the narrow content analytic. The Feynman example helps to focus attention on the real issue, namely can statements of the inferences that are constitutive of inferential roles involve true premises and *false* conclusions? Or, alternatively, must these narrow contents always be hyper–analytic? The issue is hard to discuss without agreement on which inferences *are* constitutive of inferential roles. However, I think I can sidestep important disagreements about what inferences are constitutive by considering the cases that Fodor and LePore regard as the inferential role semanticists' last refuge, traditional lexical analyticities, including those that involve natural kind terms, for example 'Cats are animals', or rather the inference from 'cat' to 'animal'. (You will recall that they hope to stick the inferential role semanticist with a commitment to such analyticities.)

Consider my Aunt Bubbles' utterance of 'Glass is a solid'. Unknown to my aunt, her utterance is *false*: glass is actually a supercooled liquid —it has the amorphous molecular structure of liquids, not the regular structure of solids. If you wait long enough, your drinking glass will "melt" into a puddle. (Really!! Ask any chemist.) There is a Twin Earth on which there is a Twin–Bubbles whose word 'solid' is anchored overwhelmingly to solids (as here), but whose 'glass' picks out a genuine solid (What she calls "glass" is not genuine glass). My aunt's 'Glass is a solid' is false, but my Twin's Aunt's utterance of the same words is true. So utterances with the same narrow content as this putative analytic truth can be both true and false: the narrow content is not hyper–analytic.

The example depends on the fact that our 'solid' can be anchored firmly to solids despite our misidentification of glass

---

[23] I will sometimes speak of the hyper–analyticity of a sentence or of a wide content or proposition, meaning the hyper–analyticity of the normally associated narrow content. The definition of 'hyper–analytic' that I gave is only a first shot. One obvious refinement —unnecesary for my purposes, is to restrict the definition to contexts in which the narrow content does yield some wide meaning or other.

as a solid. Here is another such example in which the mistake occurs on Twin Earth rather than here. Consider a Twin earth in which 'water' is used as here to refer to H_2O, but where water is very rare, most of the substances referred to as 'liquid' being slippery granular solids (Stephen White's (1982) example). 'Water is a liquid' said by them is false, although the same narrow content attaches to a truth here on earth.

I have been arguing against a hyper–analyticity for narrow meaning which a narrow meaning has if it must always yield a true wide meaning in a context. At least some narrow meanings are not hyper–analytic, as the 'Water is a liquid' and 'Glass is a solid' examples illustrate. Though these examples don't prove that there aren't other hyper–analyticities, the central role in inference enjoyed by "class–inclusion" inferences suggests that no inferential role semantics could be built around hyper–analyticity.

A different analog of analyticity for narrow meanings is what I have been calling constitutivity.[24] The inference from 'water' to 'liquid' is part of what is constitutive of the inferential role of 'water'. But constitutivity is not the kind of analog of analyticity that will do Fodor and LePore any good, since constitutive inferences can be false, as I pointed out above, and their (imagined) appeal to analyticity in the beginning of this section required a connection between analyticity and truth.

Earlier, I said I would argue against the Plausible–Sounding Principle that inferences that are included in (and therefor part of what is constitutive of) inferential roles are thereby analytic. And I said that the same argument would suffice to show the error of Fodor's and LePore's claim that compositional inferences are analytic, and Boghossian's claim that commitment to determinate meaning requires a commitment to determinate analyticity. The arguments just presented are intended to fill the bill. Narrow meanings don't determine any truth conditions at all, so Boghossian's argument

[24] See Katz (1990), p. 192, and Katz, forthcoming, for an account of constitutivity (though he doesn't call it that) that applies outside the context of inferential role semantics.

is stopped short —at least for narrow meaning. And inferences among narrow meanings can't be analytic, so the Plausible–Sounding Principle is disabled and compositionality doesn't entail analyticity. I also considered an imagined reply involving the substitution of hyper–analyticity for analyticity in Fodor's and LePore's and Boghossian's argument, and I argued that this move would gain them nothing.

Fodor and LePore also argue that since the inferential role semanticist must accept analyticity as the criterion for inclusion in inferential roles, he must give up on naturalizing those roles, that is, he must give up on characterizing these roles causally (p. 27). Without suggesting optimism about the prospects for causal characterization of inferential roles, I want to point out that once we see that the inferential role semanticist can think of inferential roles narrowly, escaping any analyticity requirement, the Fodor–LePore objection to naturalization loses its force. They also argue (in 1992a) that inferential role semantics, holism, compositionality and no analytic/synthetic distinction are an inconsistent quadruple. But this is not so for holism of *narrow* contents. Fodor and LePore also argue that the need for an analyticity requirement deprives the inferential role semanticist of his favorite argument for holism. ("...the present situation is rife with ironies...inferential role semantics doesn't *have* any holistic implications after all..." (p. 31). But with the disappearance of the analyticity requirement, that argument also dissolves.

Of course, this line of reply carries with it a promissory note of gargantuan proportions —to fill in the sketchy remarks that are often made about inferential role theories. I don't know if this promissory note will ever be cashed, but surely we should not believe arguments that depend on its uncashability —such as the Fodor–LePore and Boghossian arguments— without an argument that it is uncashable.

One final comment on the Boghossian argument: I am often asked why narrow meaning facts are supposed to be *meaning* facts and why narrow content facts are supposed to be *content* facts. As I mentioned, meaning and content have two different homes, one in the realm of communication, the other in the realm of psychological explanation. The claim of narrow meaning to be meaning rests mainly in the latter home. This is why theories that distinguish two components of meaning (e.g. sense and reference) typically appeal to examples that involve same reference/different cognitive significance. It is these differences in cognitive significance that make for differences in psychological explanation. Further, narrow contents and meaning determine truth conditions *relative to contexts*, so there is an intimate relation between narrow meaning and content on the one hand, and truth on the other. That's enough for me.

Note that I have not disuputed Boghossian's argument as it applies to wide contents. So, given that I accept wide contents, if Boghossian is right, I will have to accept facts of the matter about analytic relations among these wide contents, so isn't my reply to Boghossian peripheral to his purposes? Perhaps, but my reply to Boghossian nonetheless plays an important role in my dispute with Fodor and LePore. If Boghossian's point were right about narrow content, then I could not defend holistic narrow contents without giving up agreement and disagreement. I do not defend holistic wide contents, so Boghossian's argument, if it is right, does not give a boost to Fodor and LePore.

Boghossian has pointed out (in conversation) that my example of the narrow content of 'Glass is a solid' could be circumvented if we restricted attention what might be called "quasi–hyper–analyticities", narrow contents that generate truths in all contexts *in which they generate the same wide content.* This maneuver may have a point for some purpose, but not for the purpose of this paper, which is to show how a holistic inferential role semantics can avoid both the Old Hat and the Fodor–LePore argument. Boghossian had argued that a commitment to determinate meaning requires a commitment to determinate analyticity. I have pointed out that a commitment to determinate narrow meaning does not require a commitment to determinate analyticity (or hyper–analyticity) involving these meanings, and this allows me to avoid the Fodor–LePore argument and the Old Hat. For these purposes, it does not matter whether a commitment to determinate meaning requires a commitment to determinate quasi–hyper–analyticity.

7 More Objections

Jerry Fodor responded to an earlier version of this paper by saying that "stereotypes" provide an obvious counterexample to the Plausible Sounding Principle. My Aunt Bubbles and her Twin share the inference from something's having the stereotype of glass to it having the stereotype of solid. Though one inference is correct, the other is not. So the reasoning that I gave about narrow contents goes through equally well with stereotypes.

The first thing to say is that Fodor and LePore *need* the Plausible Sounding Principle, so to the extent that its refutation is obvious, too bad for them. Note that it is no trick to refute the Plausible Sounding Principle (or Boghossian's argument for that matter) by adopting a *plainly inadequate* theory of wide meaning. If wide meanings are held to be *mental images*, for example, both the Plausible Sounding Principle and Boghossian's argument will fail. As Putnam (1975) noted, meanings in any sense in which they determine reference cannot be e.g. images or stereotypes, since Twins can have the same 'water'–images, but different 'water'–references. If meaning determines reference, then different reference requires different meanings; hence *meanings ain't in the head*, as Putnam emphasizes, and the Twin Earth word 'water' cannot be translated by our 'water'. Stereotypes and images can be immediately disqualified as *wide* meanings, though they can be taken as candidate *narrow* meanings, in which case the Fodor point under discussion reduces to my point about narrow meaning.

Jerry Fodor has also argued (in conversation) that my 'Glass is a solid' argument does not challenge the hyper–analytic status of what he regards as compositional analyticities such as 'Brown cows are cows'.

To see what the point might be, let us go over the role of hyper–ananlyticity in the argument I gave. Fodor and LePore make a number of claims involving analyticity. I pointed out that a plausible inferential role semantics deals in narrow contents which are never true and therefor never analytic, and further, that narrow contents of the putative "lexical analyticities" that Fodor and LePore try to stick the inferential role semanticist with are not always narrow contents of analyticities —that is, they are not hyper–analytic— so hyper–analyticity will not substitute for analyticity in the Fodor–LePore argument. How could the putative hyper–analyticity of 'Brown cows are cows' resuscitate the Fodor–LePore argument?

Recall how the Fodor–LePore argument goes: The inferential role semanticist is not supposed to be able to include the inference from 'rattling snake' to 'dangerous' in the inferential role for 'snake' because, according to Fodor and LePore, that would make this inference analytic, which, they say, it is not. Then Fodor and LePore say that the inferential role semanticist has to fall back on analyticity as a criterion for inclusion of inferences in his inferential roles. So there are two distinct appeals to analyticity: (a) If all inferences are included in inferential roles, then all inferences would be analytic; but some are not. (b) Analyticity is the only fall–back criterion.

Can the putative analyticity of 'Brown cows are cows' save a version of either (a) or (b) with 'hyper–analyticity' substituted for 'analyticity'? The point against (b) is easy to see. These putative "quasi–syntactic hyper–analyticities" are not something that any serious inferential role semantics could fall back on, because there aren't enough of them. Such an inferential role semantics could make such semantic distinctions such as that between 'brown' and 'fake', but not between 'brown' and 'blue'. If "quasi–syntactic" inferences (in which a word is repeated) are the only hyper–analyticities, then an inferential role semantics that restricted itself to hyper–analyticities would be a non–starter.

As for (a), I pointed out earlier that the claim that the inference from 'rattling snake' to 'dangerous' is not *really* analytic is question–begging in the context of a dispute about holism —whether being dangerous if rattling is genuinely part of the meaning of 'snake'. The putative hyper–analyticity of 'Brown cows are cows' is irrelevant to this issue. Certainly it gives us no reason to think that including the inference from 'rattling snake' to 'dangerous' in the inferential role for 'snake' will make this inference hyper–analytic. I also pointed out that the inference to analyticity depended on the Plausible Sounding Principle, which is disarmed by the 'Glass is a solid' argument. The Plausible Sounding Principle is thus falsified —even a "hyper–analyticity" version of it— despite the possibility that there are *some* cases of inferences that are constitutive of inferential roles that are hyper–analytic. A principle is falsified by exceptions even if there are some cases that comport with it.

Finally, Burgean arguments can be brought to bear on the putative hyper–analyticity of 'Brown cows are cows'. Consider a Twin of me whose language community uses 'brown' to mean *fake*. His deference to his language community will lead *him* to regard his utterance of 'Brown cows are cows' as false (let us say) when he learns the true meaning of 'brown' in his language community. Shouldn't we agree? Objections may appeal to "idiolect" meaning, but given my Twin's insistence that what he said was false rather than idiosyncratic and true, the idiolect move can be fended off in the same way that Burge fends off the "tharthritis" reply.

A quite different sort of objection is suggested by parts of Chapter 5 of Fodor's and LePore's (1992), and by Fodor (1987), Ch 3. I have the impression that Fodor and LePore think that the two–factor approach that I endorse is doomed for a reason not yet mentioned, and so they may feel justified in letting their argument presuppose its falsity. Their objection to the two factor theory is that there is nothing to keep the two factors stuck together. Why, they ask,can't you have a sentence that has the inferential role of 'Water is wet,' but is true iff 4 is a prime number? I admit to being puzzled about why this is supposed to be an objection to inferential role theory.[25] What puzzles me about the objection is that though there is an interesting issue here, I don't see why there is supposed to be more of a *problem* for the two factor theorist than for anyone else. The issue arises for anyone who

---

[25] My response to their question is that I *don't know* whether there could be a Twin of mine whose thought or utterance of 'Water is wet' shared the inferential role of my thought or utterance of those words even though his embedding in his world was such that its truth conditions had to do with numbers. I suspect that if you run with Burge's arguments about deference to the language community, you could make a case for this possibility.

accepts the *existence* of inferential roles and of truth conditions, whether or not they think there is any plausibility in an inferential role semantics. Fodor and LePore never suggest that either inferential roles or truth conditions don't *exist*; indeed, their arguments presuppose that inferential roles (and truth conditions) do exist, so the issue they raise is as much an issue for them as it is for the inferential role semanticist.[26]

8 Addendum

Having seen a draft of Fodor's and Lepore's reply, I cannot resist adding a short note: You will recall the distinction mentioned above between semantic compositionality and psychosemantic compositionality. The former consists in the existence of a function from the semantic values of parts to wholes; the latter has to do with how people represent and process these semantic values, and thus it is the latter that can explain productivity and systematicity. (Systematicity is the fact that someone who can think the thought that John loves Mary can also think the thought that Mary loves John; this is a psychological fact, having to do with how human thought works and is not to be explained by a semantic theory; there are possible thinkers and speakers whose language has a semantics like ours but whose thought is not as systematic as ours.)

Inferential role semantics (like possible worlds semantics) is compositional in the former sense —it specifies the relevant function— though it has little to say about the latter. In both the book and the article, Fodor and LePore talk mainly in terms of the former —semantic compositionality. Thus in the book's Glossary, they define 'compositionality' as follows:

---

[26] An earlier version of this paper was given in response to Fodor and LePore at the meeting of the Society for Philosophy and Psychology, June 1991, and the July, 1991 meeting of SOFIA in Salamanca. I am grateful to Paul Boghossian, Martin Davies, Paul Horwich, Jerry Katz, Eric Lormand and Robert Stalnaker for comments on an earlier draft, and also to Jerry and Ernie for our lively discussions. I am also grateful to Earl Conee for sending me his reply to Fodor and LePore, which takes a very different line from mine.

"A language is compositional iff (idioms aside) the meaning of its syntactically complex expressions is a function of their syntactic structures together with the meanings of their syntactic constituents". However, it has now become clear (from Fodor's and LePore's reply) that it is the latter psychosemantic sense of 'compositionality' that they have in mind, and so passages such as the definition just quoted should be ignored. Given this change in view, the flaw in their main argument can be seen to be their claim that compositionality is non–negotiable. It is no responsibility of a purely semantic theory to say how people represent or process the meanings of elements of their language. Indeed, part of the role of a purely semantic theory is to provide a framework for posing questions about processing. (I am indebted to Bob Stalnaker here.) So, for example, on a possible worlds account, any candidate for how people represent predicates will have to determine a mapping of worlds onto extensions.

A last remark about Boghossian's response: just as narrow meanings cannnot be true solely in virtue of meaning, for much the same reason they cannot be asserted solely in virtue of meaning either. To assert something is to commit oneself to its truth, and so what can't be true cannot be asserted at all. Further, the class of narrow contents at issue in my paper aren't parts of wide contents that are assertible solely in virtue of meaning. My narrow content of 'water is a liquid' can be part of a false wide content on an appropriate Twin Earth, and so my Twin's assertion of it can be correctly challenged. It is part of our commitment in the use of natural kind terms that the world has a role to play in determining the truth value of what we say, and this is reflected in our implicit recognition that the world can always thwart our assertions. Nonetheless, I am willing to allow that there are some hyperanalyticities, and they will also be hyperassertible.

9 BIBLIOGRAPHY

Block, N. 1986 "Advertisement for a Semantics for Psychology", in P. French, et. al., *Midwest Studies in Philosophy* X. University of Minnesota Press: Minneapolis.

Block, N. 1987. "Functional Role and Truth Conditions", in *Proceedings of the Aristotelian Society*, Supplementary Volume LXI. Blackwell: Oxford.

Block, N. 1991. "What Narrow Content is Not", in B. Loewer and G. Rey, *Fodor and His Critics*. Blackwell: Oxford.

Boghossian, P. forthcoming. "Analyticity", in *The Blackwell Companion to Philosophy of Language*. Blackwell: Oxford.

Devitt, M. 1990. "A Narrow Representational Theory of Mind", in W. Lycan (ed.), *Mind and Cognition: A Reader*. Blackwell: Oxford.

Dretske, F., 1988. *Explaining Behavior*. MIT: Cambridge.

Field, H. 1977. "Logic, Meaning and Conceptual Role", *Journal of Philosophy* 74, pp 379-409.

Fodor, J. 1987. *Psychosemantics*. MIT: Cambridge.

Fodor, J. 1990. "Substitution Arguments and the Individuation of Beliefs", in *Meaning and Method: Essays in Honor of Hilary Putnam*, ed. G. Boolos, Cambridge University Press, Cambridge, 1990, pp. 63-78.

Fodor, J. and LePore, E., 1992a. *Holism: A Shopper's Guide*. Blackwell: Oxford.

Fodor, J. and LePore, E., 1992b. "Why Meaning (Probably) Isn't Conceptual Role", *Mind and Language* XX: XX.

Harman, G. 1982. "Conceptual Role Semantics", *Notre Dame Journal of Formal Logic* 23: 242-56.

Horwich, P. 1990. *Truth*. Blackwell: Oxford.

Horwich, P. 1992. "Quine versus Chomsky on the Analytic–Synthetic Distinction", in *Proceedings of the Aristotelian Society, 1991-1992*. Blackwell: Oxford.

Horwich, P. forthcoming. "Scientific Conceptions of Language and Their Philosophical Import", SOFIA conference volume. Ridgeview: Atascadero.

Katz, J. 1972 *Semantic Theory*. Harper and Row: New York.

Katz, J. 1990 *The Metaphysics of Meaning*. MIT Press: Cambridge.

Katz, J. forthcoming. "The New Intensionalism".

Kripke, S. 1979. "A Puzzle about Belief", in A. Margalit, ed. *Meaning and Use*, Dordrecht: Reidel.

Loar, B. 1987. "Social Content and Psychological Content", in *contents of Thought: Proceedings of the 1985 Oberlin Coloquium in Philosophy*, eds R. Grimm and D. Merrill. University of Arizona Press: Tucson.

Putnam, H. 1975. "The Meaning of 'Meaning'", in *Mind, Language and Reality: Philosophical Papers*, vol 2. Cambridge University Press: Cambridge.

White, Stephen, 1982. "Partial Character and the Language of Thought", *Pacific Philosophical Quarterly* 63: 347-365. Reprinted in White's *The Unity of the Self*, MIT: Cambridge, 1991.

3

PHILOSOPHICAL ISSUES, 3
Science and Knowledge, 1993

Does an Inferential Role Semantics Rest Upon a Mistake?

Paul A. Boghossian[1]

1 Introduction

Let's suppose that we think in a language of thought and that there are causal facts of the following form: the appearance in O's belief box of a sentence S_1 has a tendency to cause the appearance therein of a sentence S_2 but not S_3. Ignoring many complications, we may describe this sort of fact as consisting in O's disposition to *infer* from S_1 to S_2, but not to S_3.

Let's call the totality of the inferences to which a sentence is capable of contributing, its *total inferential role*. A sub–sentential constituent's total inferential role can then be defined accordingly, as consisting in the contribution it makes to the total inferential role of the sentences in which it appears.

[1]I am grateful to Ned Block, Jennifer Church, Stephen Schiffer and the members of my seminar on Analyticity at the University of Michigan, for helpful discussion and/or comments.

Against this rough and ready background, an *inferential role semantics* is just the view that there is some construct out of an expression's total inferential role that constitutes its meaning what it does. Let us call this construct an expression's *meaning-constituting inferential role*, or MIR, for short.[2]

As no one will care to deny, at the present time an inferential role semantics (hereafter, IRS) is no more than a twinkle in the eyes of its advocates. To this day, no one has proposed a version that is detailed enough to be called a "theory"; in particular, no one has even come close to providing a plausible specification of the constructs that are to constitute MIRs.

2 Fodor and Lepore's Argument

According to Fodor and Lepore, however, no one should even bother trying. For F & L claim to have on hand a powerful argument to the effect that *any* attempt to specify an expression's MIR will founder on one of two insuperable difficulties. The crux of their argument is neatly summarized in the following passage:

> We can now say pretty exactly what our problem is: You can't identify meanings with inferential roles *tout court*, since, unlike meanings, inferential roles *tout court* aren't compositional. You *can* identify meanings with roles in *analytic* inferences, however, because analytic inferences *are* compositional. But, of course, the cost of identifying meanings with roles in analytic inferences is buying into the analytic/synthetic distinction.... But, these days, practically everybody thinks that the analytic/synthetic distinction is unprincipled. (p. 27)

---

[2] Although Fodor and Lepore's title suggests that their paper is to be about *conceptual* role theories in general, rather than just about *inferential* role theories —inferential role being the purely linguistic aspect of an expression's overall causal role, a role that will typically include relations to non-linguistic stimuli or objects— their argument concentrates exclusively on the inferential part. For the sake of simplicity, therefore, I have framed the discussion as concerning inferential role theories only.

The *structure* of their argument seems pretty clear (though, as we shall see, its *content* is somewhat less transparent). An inferential role semanticist has to specify what is to constitute the MIR of an expression. And here he would appear to have only two choices: to identify the MIR of an expression with its total inferential role (what the authors refer to as its "inferential role *tout court*"); or to identify its MIR with some proper subset of that total role. However, the authors contend, each of these options faces decisive objections.

Identifying the MIR of an expression with its total role leads to a theory of meaning that violates compositionality; and that is unacceptable. That leaves open the option of identifying it with some proper subset of that total role. But which subset should that be? The only plausible suggestion —indeed, the only suggestion that respects compositionality— is the identification of an expression's MIR with its role in *analytic* inference. However, and unfortunately, "these days practically everybody thinks that the analytic/synthetic distinction is unprincipled".

And so, the authors conclude, there is probably no satisfactory way to implement an inferential role semantics. Perhaps it is time that the philosophy of language reassessedthe plausibility of an old–fashioned word–world semantics.

Now, I am not going to discuss F & L's argument for the first horn of their dilemma. I do not believe that a holistic version of an IRS —the version that identifies the MIR of an expression with its *total* role— violates compositionality. But since my reasons here mostly recapitulate Ned Block's, I will simply refer the reader to his contribution to this symposium.

However, given the many *other* problems with which a holistic version is beset —not the least being the well–known problems concerning change of mind, translation, and subsumption under intentional laws—considerable interest attaches to F & L's argument for their second horn. In my view —and *pace* Ned Block— the only plausible version of an inferential role semantics is a non–holistic version. It is important, therefore, to assess the effectiveness of F & L's argument against that option.

What, precisely, is that argument? As we saw, it consists in essence of the following:

(1) The only plausible version of a non–holistic IRS identifies an expression's MIR with its role in analytic inference.

(2) The analytic/synthetic distinction is "unprincipled".

Therefore,

(3) There can be no plausible version of a non–holistic IRS.

3 The First Premise

Let's begin by looking at the first premise There are in fact *two* ways of understanding (1), depending on how the phrase "identifies an expression's MIR with its role in analytic inference" is understood. On the one hand, the phrase might simply mean: *identifies an expression's MIR with its role in analytic inference, under that very description.* On the other hand, it might bear the following more complicated interpretation: *identifies an expression's MIR with its role in inferences which possess property P, where P is necessarily equivalent to the property of being analytic.* It is important, in the present connection, to distinguish between these two readings, and the distinct versions of an IRS that they determine.

Now, it is one of the more curious aspects of F & L's paper that they seem to intend that (1) be understood under its *simple*, rather than its complicated, interpretation. That is, the version of a non–holistic inferential role semantics that they take to be most plausible, and against which their argument is primarily directed, is a version that identifies an expression's MIR with its role in analytic inference, *under that very description.*

That this is the way they are construing the view under discussion is made very clear by what they have to say about it. Thus:

> The first thing to say about this new suggestion is that the threat of circularity is now very close to the surface. It is proposed that we reconcile New Testament Semantics with the compositionality of meaning by identifying the meaning of an expression with

> its role in analytic inferences. But the difference between analytic inferences and inferences tout court is just that the validity of the former is guaranteed *by the meanings* of their constituent expressions. (p. 26)

Furthermore:

> Notice also that the naturalizability of inferential role semantics is jeopardized by the present proposal. A lot of the attraction of identifying meaning with inferential role lies in the thought that the inferential role of an expression might in turn be identified with *causal* role, thereby conceivably providing the basis for a naturalistic solution to Brentano's problem.... But, barring proposals for a causal theory of analyticity, this tactic is unavailable to the philosopher who identifies meaning with the role of an expression in *analytic* inference. (p. 26-27)

As these passages make amply clear, F & L intend their argument to be directed against an IRS that proposes to specify (some expression) *e*'s MIR by saying that it consists in: the set of *e*–involving inferences whose validity is secured solely by the meanings of the participating expressions. No other hypothesis could explain their remarks. *Only* an IRS that employs a *non–reductive* notion of analyticity in its specification of MIRs faces a "threat of circularity" or a "fear that the naturalizability of semantics is [thereby] jeopardized". And that is the only version of an IRS that they describe.

The trouble, however, is that I don't see how a viable IRS could *conceivably* assume the form that F & L describe. For I would have thought that with the sort of view they have in mind the "threat of circularity" is not merely "very close to the surface" but right smack on top of it for all to see; and that, as such, it simply is not a serious option. In other words, it seems to me that regardless of the ultimate cogency of the description 'analytic inference,' it just *isn't* open to an IRS theorist to specify an expression's MIR by the *mere* use of that very description —by simply saying that an expression's meaning is determined by its role in inferences that are valid by virtue of its meaning. If an IRS theorist wishes to identify an expression's MIR with its role in analytic inference, he must identify it with its role in analytic inference *reductively* construed.

F & L may have been misled into thinking otherwise by their insistence that, for the purposes of the present discussion, they do not wish to presuppose "physicalism".

> We remark in passing that none of this is to be construed as an attempt to legislate physicalism. For present purposes we're content that semantic properties should be, for example, irreducibly intentional, or irreducibly epistemological, or irreducibly teleological. But we take it to be not on the cards that they are irreducibly semantic. In short, we don't care whether semantic properties supervene on something that is *physical* just as long as they supervene on something other than themselves. (p. 17)

Putting aside objections to the slide from talk of reduction to talk of supervenience, the trouble is that any IRS that offers nothing more than an unreduced notion of analyticity in its definition of MIRs —as F & L's targeted IRS proposes to do— would not have provided *any* account of what, apart from the semantic, semantic properties are supposed to be identical with or supervenient upon. It's one thing to say that an IRS is entitled to use an unreduced notion of *inference*; it's quite another to say that it is entitled to use an unreduced notion of *validity by virtue of meaning*. The former doesn't necessarily threaten the informativeness of the relation it is supposedly trying to specify; the latter most certainly does.

I am inclined to think, therefore, that there is a serious question whether theoretical space even *allows* for the sort of view that F & L designate their most worthy opposition. For present purposes, however, I intend to waive this point. After all, a reductive version of an analyticity–based IRS survives it; and the soundness of F & L's second premise —and hence of the anti–IRS argument that they base on it— doesn't turn on whether the implicated notion of analyticity is construed reductively or not.

What, then, about the substance of their claim that the "most plausible" version of a non–holistic IRS would identify *e*'s MIR with the set of analytic inferences involving it? For reasons that will emerge below (in my discussion of the "Plausible–Sounding Principle"), I do not believe that F & L provide any support for this claim. But here again I propose to be concessive and to grant it to them. My worries about their argument are concentrated on their second premise, rather than on the first.

4 The Second Premise

So let us turn to looking at that second premise. To begin with, what does it mean to say that the analytic/synthetic distinction is "unprincipled?" The authors provide the following explicit explanation:

> This means not that the distinction is *vague* —what empirical distinction isn't— but that there aren't any expressions that are true or false solely by virtue of what they mean. (p. 20)...
>
> ...the argument that there is no analytic/synthetic distinction is the argument that there *aren't any* analytic inferences. (p. 32)

Notice that according to the explanation provided, the notion of an 'analytic inference' is a purely *semantical* idea: to say that an inference is analytic is just to say that facts about its meaning suffice to fix its validity. In particular, the notion as defined appears to contain no *epistemological* component, mandating that analytic inferences be knowable in a certain way, for instance *a priori*.

F & L's idea, then, seems to be this: Quine showed (or at a minimum gave very good reason to believe) that there aren't any inferences(/sentences) that are valid(/true) solely by virtue of what they mean; and this is inconsistent with the claim that the meaning of an expression is determined by its role in analytic inference.[3]

Well, I am prepared to concede the second part of this claim, and hence the *validity* of the argument outlined. If it were true that the extension of the predicate 'analytic

---

[3] Notice that, as I said above, this objection doesn't depend on whether the inferential role semantics is thought of as leaning on a reduced or unreduced notion of analytic inference. If there simply aren't *any* analytic inferences, then there aren't any reduced analytic inferences either.

inference' were empty, then the IRS theorist targeted by F & L *would* be in serious trouble. Obviously. For this theorist seeks to explicate e's meaning in terms of its meaning–constituting inferential role; and he seeks to specify that role by saying that it consists in the set of e–involving analytic inferences (construed either reductively or non–reductively). If, however, there are no analytic inferences, then every expression gets assigned the same 'null' meaning, and that is clearly unacceptable.

The only snag with this argument is that I don't think it can plausibly be claimed that we have been given —by Quine or by anyone else— good reason to believe that there are no analytic inferences. At a minimum —and more importantly for present purposes— F & L can't claim that we have.

To see why, we need to look briefly at a train of thought that I have developed at greater length elsewhere (forthcoming). Let us suppose —what is surely common ground between the principal disputants in *this* debate— that meaning realism is true and hence that there are determinate facts about what means what. Then, condensing considerably, there are bound to be facts about whether two expressions do or do not mean the same, and, hence, about whether the inference from

This is e

to

This is e'

is or is not analytic (valid by virtue of meaning alone).[4]

At a minimum, then, the meaning–realist must suppose that 'analytic inference' expresses a coherent, determinate

[4] Throughout this paper, the phrases "valid by virtue of meaning alone" and "true by virtue of meaning alone" are to be understood as meaning valid (or true) by virtue of meaning *and logic* alone. In "Analyticity" I distinguish between this 'impure' idea of analyticity and the 'pure' version that construes even the truths of logic as non–trivially true by virtue of meaning alone. This latter *positivist* notion of analyticity is not at issue here. For present purposes, my claim is only that any meaning–realist is at least committed to the existence of impure analyticities. For further discussion see my forthcoming.

property in good standing. Moreover, it is hard to see how he could plausibly avoid believing that it has a non–empty extension, that there actually are inferences that instantiate the property of being valid by virtue of meaning alone. For it is hard to see how it could plausibly turn out that, although the property of meaning identity is coherent and determinate, no two expressions could, either in point of principle, or in point of fact, instantiate it. (For the arguments, again see my forthcoming.)

Making a long story very short, then, I don't see that any meaning–realist can accept the claim that there are no inferences valid by virtue of meaning alone. *A fortiori*, I don't see that F & L can.

5 Block's Objection

Ned Block has objected to this train of thought (see his contribution to this symposium). He accepts it, he says, for *wide* meanings, but not for *narrow* meanings. Why? According to Block, the point is,

> that narrow contents are *never* analytic. They aren't even the sort of thing that can be analytic. To see this, note that narrow contents are never *true*, and hence have no truth conditions. Analyticity requires truth —truth in all conditions, so something that can't be true can't be analytic either. (p. 61)
>
> Narrow meanings don't determine truth conditions at all, so Boghossian's argument is stopped short —at least for narrow meanings. (p. 63-64)

The point, if true, would be particularly *a propos* in the present context since, as Block observes, inferential role semantics are typically proposed as theories of narrow rather than of wide content. But is it true?

Not, I think, in any sense that matters to the issues presently at stake. It goes practically without saying that if we stipulate that what's in play is a notion of meaning that is not truth–conditional, then, relative to this stipulation, there won't be any sentences that are true purely by virtue of meaning, nor any inferences that are valid purely by virtue of meaning, truth and validity being —unavoidably— truth–theoretic properties. This would be, however, an entirely trivial consequence of the initial stipulation.[5]

It follows, therefore, that if we allow for a kind of meaning that is not truth–theoretic in nature, that a mere commitment to determinate meanings will not by itself entrain a commitment to analyticity —defined as truth (or validity) by virtue of meaning.

But that is irrelevant to the essential core of the argument I was concerned to develop and defend. The important point for my purposes —and the one that really engages Quine's critique of analyticity— is the observation that a believer in determinate meanings has no choice but to recognize that certain sentences or inferences will *hold* —to use as neutral a term as possible— merely as a result of the meanings of their constituent expressions. Whether this notion of "holding" is cashed in terms of truth or assertibility or whatever —and, hence, whether the associated notion of analyticity is thought of as 'truth by virtue of meaning alone' or 'assertibility by virtue of meaning alone'— is immaterial from the standpoint of the argument. How it is cashed will depend, obviously, on which notion of meaning is in play and what conception of cognitive virtue it defines.

What Quine was concerned to deny —and my argument to affirm— was that there could be cases where the obtaining of the relevant cognitive virtue could be attributed solely to facts about the underlying meaning. Quine found it natural to think of that cognitive virtue as truth, and so framed the debate as an issue about the possibility of truth (or validity) by virtue of meaning alone. Clearly, however, had he been thinking in terms of assertibility, say, instead of truth, he would have been equally unhappy with the idea of 'assertibility purely by virtue of meaning alone.' What is essential to

[5] And certainly not the sort of deep rejection of analyticity that Quine was after. Quine is very careful in "Two Dogmas" to make clear that his rejection of analyticity is supposed to hold for a truth–conditional notion of meaning. (I don't mean to suggest that Block isn't aware of this; his point isn't meant to be a defense of Quine, but merely an attempt to stem the generality of my argument. I do not believe that it succeeds in that task.)

Quine is the claim that cognitive virtue, however defined, is never cogently seen as a consequence of meaning alone. And that is what I have argued a meaning realist can never accept.

F & L emulate Quine in thinking of meaning and, hence, of analyticity, truth–conditionally; in discussing their argument, I have followed suit. However, none of the crucial points depend upon this; they can all be recast, if necessary, in terms of a non–truth–conditional notion of narrow meaning. Since nothing hangs on it, I shall continue to use F & L's truth–conditional framework, leaving the generalization to other frameworks implicit.

6 The Plausible–Sounding Principle

Before we go back to the main line of argument, I have another bone to pick with Block. Block formulates a thesis he calls the "Plausible–Sounding Principle":

> Inferences that are part of [MIRs] must be regarded by the inferential role theorist as analytic. For these inferences are what are taken to constitute meaning, and inferences that constitute meaning are analytic. (p. 51)

He goes on to claim (a) that this principle is presupposed by F & L's argument, (b) that it is an instance of the more general connection claimed by me between meaning–realism and analyticity and, therefore, (c) that it is true for wide meanings but false for narrow meanings for the reasons that we had occasion to review and reject above.

I won't take a stand here on whether F & L really do require the PSP. But it is important to see that neither of Block's other two claims is true. The PSP is neither an instance of the general connection between meaning–realism and analyticity that I have outlined nor is it independently true, even for wide meanings.

It can seem as if the following argument simply clinches the PSP.[6] If I say that the inference from 'rattling' to 'dan-

[6] If truth be told, I myself argued in this way in an earlier draft of this paper.

gerous' is part of the MIRs of 'rattling' and dangerous', then I am saying that it is *necessary* for 'rattling' and 'dangerous' to mean what they do that that inference be valid. And if that inference's being valid is necessary for the expressions to mean what they do, then the expressions' meaning what they do is *sufficient* for the validity of the inference —which is just to say that the inference is analytic.

The fallacy here consists in conflating *being regarded as valid* with *being valid*. To say that the inference from 'rattling' to 'dangerous' is part of the MIRs of 'rattling' and 'dangerous' is only to say that someone has to be prepared to infer from 'rattling' to 'dangerous', if he is to mean by those words what they in fact mean. At best, this commits one to saying that someone has to be prepared to *regard* that inference as valid; it makes no commitment whatever on whether that inference *is* valid, let alone whether it is valid by virtue of meaning alone.

The point here is a general one. A theory of meaning may well end up deciding (and many theories of an 'Interpretational' sort have decided) that if someone doesn't believe certain obvious, pervasive *empirical* truths about dogs, then they can't mean *dog* by 'dog'. I am not saying that a theory of meaning is well–advised to decide this, just that it can do so without fear of logical inconsistency. To say that someone must regard a certain sentence as true, if he is to mean this by that, is not the same as saying that the sentence *is* true. *A fortiori*, it is not to say that the sentence is true by virtue of meaning alone.

Just so, an IRS theorist may end up ruling that no one counts as meaning *dog* by 'dog' unless that person is prepared to infer from 'x is a dog' to 'x is an animal'. But it doesn't *follow* that that inference is analytic. It may well be part of an inferential conception of meaning that all sorts of inferences that aren't valid by virtue of meaning alone are nevertheless parts of MIRs, nevertheless *constitutive* of meaning.

Notice that this point, too, is independent of the narrow/wide distinction. Even if we operated within a non–truth–conditional framework —and so thought of inferences as *holding*, say, rather than as valid— we would still need to recognize a distinction between *regarding* an inference as holding and its actually holding. Being constitutive of meaning is one thing, being analytic another.

In sum, then, I reject both the claim that the PSP is a special case of the internal connection between meaning–realism and analyticity and, indeed, the PSP itself.

7 F & L Continued

Returning to the main line of argument, let us take stock. F & L have argued that an IRS can't identify the MIR of an expression with its role in analytic inference, because the analytic/synthetic distinction is "unprincipled". According to their official explanations, for an inference to be analytic is for it to have a purely semantical property: validity by virtue of meaning alone. And for the a/s distinction to be unprincipled is for there to be no inferences with such a semantical property, no inferences that are analytic in the purely semantical sense.

As against this, I have claimed that no meaning realist —*a fortiori*, not F & L— can agree that there will be no analytic inferences in the semantical sense. So their official argument against the possibility of a non–holistic IRS fails.

At this point it might have been appropriate for me to elaborate on the entailment I have claimed; and I would gladly have done so were it not for the fact that —curiously enough— F & L give every indication of accepting it! Thus, at p. 27, footnote 4, they say:

> There are causal theories of meaning around.... And like any account of meaning, each of them implies a corresponding notion of analyticity.

Their subscription to the entailment is further fleshed out in the book from which the article currently under discussion is drawn:

> [I]magine somebody who holds a reductionist theory of meaning; for example, somebody like Skinner, who holds that for "dog" to mean *dog* in a certain speaker's mouth is for the speaker to

> have the habit of uttering "dog" when there are dogs around.... Look, someone might say, if Skinner were right, then you could have a situation in which a speaker has two responses ("dog" and "shmog", as it might be) that are conditioned to *exactly the same stimuli.* But then it would follow that these responses would be *synonymous* for that speaker. So then the following sentence would be *analytic* in the speaker's language.... "Whatever is a dog is a smog". But Quine showed that there are no such things as synonyms or analytic sentences. So Skinner's semantics must be wrong. A priori! In fact, *all* semantic theories must be wrong, a priori, except for the nihilistic theory which says that there are no semantic properties.(Fodor and Lepore, 1992, pp. 56-57)

F & L rightly point out —just as I do in "Analyticity"— that this pattern of reasoning is a *reductio* of the claim that there can be no analytic inferences in the purely semantical sense.

Their official exposition notwithstanding, then, their argument against the targeted IRS theorist can't after all be that it is committed to the existence of inferences that are valid purely by virtue of meaning; by their own admission, every theory of meaning is, including the Skinner–style informational theory that they are concerned to promote. So *what* is their argument?

It's hard to find a different argument for the same conclusion in the article that is the subject of this symposium. But in their book, F & L provide the following alternative explanation of what Quine may be credited with having shown:

> We take it that, strictly speaking, Quine in "Two Dogmas" did *not* show, or even argue, that there are no analytic truths.

"He did argue", they continue in a footnote,

> plausibly in our view, that you can't reconstruct analyticity by appeal to aprioricity; that is, that if there *are* analytic truths, we don't know them a priori. (p. 57)[7]

---

[7] But compare p. 25 of the same book:

> When Quine says, "No a/s" he presumably means "No analytic sentences".

It is unfortunate that the argument of this interesting book is marred by this sort of vacillation about what "analytic" is supposed to mean.

So maybe what F & L really intended to say is that a non-holistic IRS is committed to *analyticities that are a priori knowable* and that *that* is the notion that has been shown to be empty by Quine.

Well, if that is the argument F & L wished to make in the paper that is the subject of this symposium, it is arguable that they should have expressed themselves quite differently; but let us put that to one side. It is a substantially different argument from the one we have just considered and is not subject to the same reply. The point is that a priori analyticity is a much more demanding notion than a plain semantical analyticity; it cannot be defended by mere appeal to meaning-realism —it simply doesn't follow from a commitment to determinate meanings that meaning equivalences will be knowable in any particular way, let alone a priori. How, if at all, facts about the relations between meanings will be knowable will depend on the detailed nature of meaning, detail on which a mere realism about meanings is silent, and which awaits resolution by a satisfactory theory of meaning.

For all that, could it reasonably be claimed, as F & L do, that Quine showed that there couldn't be any a priori analyticities, so that no satisfactory theory of meaning could afford to postulate them? In my opinion, not at all (here again see my discussion in "Analyticity").

However, I don't need to argue the point here because it is completely irrelevant to the issue at hand. The fact is that *it is simply not true that an inferential role semantics is committed, by its very nature, to reconstructing meanings out of inferences whose validity is knowable a priori.* So even if we were to grant that Quine had shown that the extension of 'a priori analytic inference' were empty —something which for independent reasons I don't believe we should be prepared to do— F & L's argument against an IRS fails.[8]

---

[8]I do not deny that there are actual IRS theorists who thought that the MIRs defined by their theories would contain only a priori knowable inferences; some of them may have even thought of that as a virtue. But F & L purport to have an argument against any sort of inferential role semantics. And that is the argument that is eluding my grasp.

To get an argument that works, relative to the concession to Quine, F & L would somehow need to show that any IRS is committed to the inferences' being a priori. But where would such a demonstration come from? The only basis I can think of would have to consist in some general argument to the effect that *any* meaning–based inference has to be a priori. Whatever the merits of such a claim, however, it is not one that F & L can appeal to, for the theory of meaning that they favor is knee–deep in analytic inferences that aren't a priori: as they themselves are fond of pointing out, causal/informational theories of meaning will generate lots of inferences whose validity is secured by meaning alone, but which will not be knowable a priori. So they can hardly believe that it is a general requirement on analytic inferences that they be a priori. If there were such a requirement, it would count against a very wide range of theories of meaning, including their own (always relative, of course, to the concession to Quine).

I conclude, therefore, that regardless of whether 'analyticity' is read as expressing pure semantical analyticity or as expressing a priori analyticity, F & L's argument discloses no new difficulty for an inferential role semantics. It remains possible, of course, that there is some *other* notion of analyticity which is such that it is both required by an IRS and effectively destroyed by Quine. But if there is such a notion, I don't know what it is. And so far as I am able to judge, neither do F & L.

8 REFERENCES

Boghossian, P. forthcoming, "Analyticity", in Crispin Wright and Bob Hale (eds.): *Blackwell's Companion to the Philosophy of Language*. Oxford: Blackwell.

Fodor, J. and Lepore, E. (1992), *Holism: A Shopper's Guide*. Oxford: Blackwell.

4

Philosophical Issues, 3
Science and Knowledge, 1993

Belief Ascription and a Paradox of Meaning

Stephen Schiffer

I shall do four things in this paper. First, I shall propose a certain theory of the semantics of belief ascriptions as being the best theory of their semantics relative to a certain assumption. Second, I shall raise three problems for this best theory. Third, I shall make what I hope is an interesting connection between the main issue addressed and the vexing question about the form that a meaning theory for a particular language must take. And fourth, I shall close with a brief look at a paradox that will by then have been generated, and an equally brief statement of the solution to it that I am inclined to favor.

The assumption in question is the widely accepted one that each natural language has a correct compositional truth theory. A compositional truth theory for a particular language L is a finitely axiomatizable theory of L that issues in a theorem of the form

An utterance of 'S' is true in L iff such and such

for each of the infinitely many truth-evaluable sentences of L. For example, a truth theory for English might contain a theorem like

> An utterance of 'She wrote it' is true in English iff the female referred to by the utterance of 'she' wrote the thing referred to by the utterance of 'it'.

If we assume that English has a correct compositional truth theory, then we may ask how a particular class of sentences is to be accommodated in such a theory. This is to ask about the kinds of semantic values that would have to be assigned to the components of those sentences and the kinds of compositional rules that would need to apply to those values in order to crank out a truth condition for each sentence of the class. Roughly speaking, this question of accommodation within a compositional truth theory is what philosophers are concerned with when they ask about the *logical form* of a class of sentences. And this is my cue to restate the first of the things I shall do —namely, to offer what I take to be the best account of the logical form of belief ascriptions. In other words, I shall try to say how belief ascriptions would have to be accommodated in a compositional truth theory on the assumption that English has one.

1 The Hidden–Indexical Theory Baldly Stated

If English has a compositional truth theory, how should belief sentences be accommodated in it? What, that is to ask, is the best theory of the logical form of belief ascriptions on the assumption that they have a logical form? The theory I shall propose in answer to this question is one I shall call the *hidden–indexical theory.* It is a theory whose essential idea must have occurred to almost anyone who has thought seriously about the semantics of belief sentences. I first discussed a version of it in "Naming and Knowing";[1] I discuss it briefly in the form I shall presently give it in "The 'Fido'–Fido Theory of Belief";[2] and a version of the theory is nicely presented

---

[1] *Midwest Studies in Philosophy,* II (1977): 28-41.
[2] *Philosophical Perspectives,* I (1987): 455-480.

and defended in Mark Crimmins' and John Perry's "The Prince and the Phone Booth: Reporting Puzzling Beliefs".[3]

The theory, in my favored version, begins with the claim that believing —the relation expressed by 'believes' in a sentence of the form 'x believes that S'— is a three–place relation, $B(x, p, m)$, holding among a believer x, a certain sort of *structured proposition* p, and a *mode of presentation* m under which x believes p. The idea is that x may believe p under one mode of presentation m, disbelieve p under a second mode of presentation m', and suspend judgment altogether under a third mode of presentation m''. Actually, I should have said that believing is a four–place relation, where the fourth place is a time at which x believes p under m, but I shall suppress the temporal reference for simplicity. Let me now explain the notions of a structured proposition and a mode of presentation before moving on to the essence of the theory, its account of the semantics of belief reports.

The sort of proposition I have in mind is a structured entity whose basic components always include at least one property or relation, and may include any sort of contingently existing object. The simplest example of such a structured proposition would be a so–called *singular proposition* such as the ordered pair $\langle$Fido, doghood$\rangle$. This proposition is true just in case Fido instantiates doghood, the property of being a dog; that is to say, just in case Fido is a dog. Interesting technical questions arise about the representation of more complex propositions, but singular propositions will serve present purposes well enough. All that really matters to the hidden–indexical theory is that the propositions we believe are structured entities composed out of the objects and properties we are pretheoretically taken to be talking about when we express our beliefs. I shall call such propositions *Kaplan propositions*, in honor of David Kaplan, who did so much to popularize them.[4]

---

[3] *The Journal of Philosophy*, LXXXVI, 12 (1989): 685-711.

[4] Let a *singular proposition* be any ordered pair $\langle\langle x_1, \ldots, x_n\rangle, \Phi_n\rangle$, where $\langle x_1, \ldots, x_n\rangle$ is an n-ary sequence of items and Φ_n is an n-ary property (brackets are customarily dropped for one-membered sequences). Such a proposition is true just in case its n-ary sequence instantiates

This brings us to modes of presentation. A mode of presentation of a *proposition* is determined by (a) the modes of presentation of the objects and properties contained in the proposition and (b) the place of those things in the proposition. As a first approximation, you might think of a mode of presentation of a proposition as an n–tuple of modes of presentation of the objects and properties contained in the proposition.

The expression 'mode of presentation', as it was introduced by Frege and as it is used today, is a technical notion, and to understand it is to appreciate the need for its introduction. Two little stories should help remind us of that need.[5]

The first is the infamous case of the morning dog and the evening dog. There is a certain dog who begs at Ralph's door every morning. Ralph feeds this dog, whom he has named 'Fido', and has grown attached to it. Ralph believes that Fido is male. There is also a certain dog who begs at Ralph's door every evening. Ralph feeds this dog, too, whom he has named 'Fi Fi', and has also grown quite attached to it. Ralph believes that Fi Fi is female, and he thinks that Fido and Fi Fi would make a really cute couple. Unbeknown to Ralph, Fido *is* Fi Fi.

The second story is that of the spurious natural kind shmoghood. Ralph came upon a race of creatures which he thought comprised a previously unencountered biological species, and he introduced the word 'shmog' to designate

its n-ary property. One way to go with Kaplan propositions is to treat them all as singular propositions. Thus, the proposition *that Fido is a dog and Pieface is a cat* might be

$$\langle\langle\langle \text{Fido, doghood}\rangle, \langle \text{Pieface, cathood}\rangle\rangle, \text{CONJ}\rangle,$$

where CONJ is that relation between two propositions that obtains iff both are true; and the proposition that some dogs bark might to a first approximation be represented as

$$\langle \text{the property of being a dog that barks, SOME}\rangle,$$

where SOME is that property that a property has iff it is instantiated. And so on.

[5]The next couple of paragraphs derive from my "The Mode–of–Presentation Problem," in C. Anderson and J. Owens, eds., *Propositional Attitudes* (Stanford: CSLI, 1990).

members of that species. "A thing shall be called a 'shmog'", Ralph said, "just in case it belongs to the species of these creatures". Unbeknown to him, however, shmoghood is doghood; Ralph had stumbled not upon a new species but upon a new race of dogs, and thus the property that 'shmog' has been introduced as standing for is none other than doghood.

Now when Ralph says "Fido is male but Fi Fi is female", he is not being irrational in holding the belief which that utterance expresses. But he could not rationally hold the belief that would be expressed by an utterance of 'Fido is and is not male'. Likewise, Ralph is not irrational in holding the two beliefs expressed by a morning utterance of 'This dog is male' and an evening utterance of 'This dog is female', even though the same dog is referred to in both utterances. But Ralph could not rationally hold the belief that would be expressed by a sincere utterance of 'This dog is and isn't male'.

Similarly, when Ralph says "Fido is a dog, not a shmog", and when he says "No dogs are shmogs", he is not being irrational in holding the beliefs those utterances express. But he would be irrational were he to hold the belief that would be expressed by an utterance of 'No dogs are dogs'.

Appeal to the notion of a mode of presentation is designed precisely to accommodate these sorts of data. The intuitive idea, definitive of the notion of a mode of presentation, may be called *Frege's Constraint.* Stated informally and in a way that prescinds from the details of any particular theorist's account of belief or of belief–ascribing sentences, Frege's Constraint has two parts. First it says that a rational person x may both believe and disbelieve that a certain thing or property y is such and such only if there are distinct modes of presentation m and m' such that x believes y to be such and such under m and disbelieves it to be such and such under m'. Then it says that there are distinct modes of presentation m and m' such that rational person x believes y to be such and such under m and disbelieves y to be such and such under m' only if x fails to realize that m and m' are modes of presentation of one and the same thing. In other words, you cannot rationally believe and disbelieve something under one and the same mode of presentation, or under modes of presentation that you realize are modes of presentation of the same thing. The notion of a mode of presentation is *functionally defined* by Frege's Constraint in that something is a mode of presentation if it plays the role defined by Frege's Constraint, and nothing can be a mode of presentation unless it plays that role. *To ask what modes of presentation are is just to ask what things play that role.* Presently I shall raise the question of what modes of presentation might be, but the hidden–indexical theory I shall now describe is so far neutral on the outcome of that question.

As applied to the paradigmatic example

[1] Ralph believes that Fido is a dog

the hidden–indexical theory says that the logical form of an utterance of this sentence may be represented as

[2] $(\exists m)(\Phi^{*}m \,\&\, B(\text{Ralph}, \langle\text{Fido}, \text{doghood}\rangle, m))$,

where Φ^{*} is an implicitly referred to and contextually–determined type of mode of presentation. By a type of mode of presentation I mean merely a property of modes of presentation; Φ^{*}, for example, might be that property that a propositional mode of presentation has when and only when it requires thinking of Fido as being the dog who appears in the morning and requires thinking of doghood as a property shared by such–and–such similar–looking creatures. The reference to a type of mode of presentation is implicit in that although the sentence requires the speaker to be referring to a type of mode of presentation whenever the sentence is uttered, there is no word in [1] which refers to that type (whence the 'hidden' in 'hidden–indexical theory'). The reference to the type of mode of presentation is "contextually determined" in that different types may be referred to on different occasions of utterance (whence the 'indexical' in 'hidden–indexical theory').

As we shall see when I stop baldly stating the theory and begin defending it, the reference is to a type of mode of presentation rather than to some particular mode of presentation because we need not be in a position to refer to an actual mode of presentation under which someone believes a proposition. But in a limiting case reference may be to an actual mode of presentation, for the "type" referred to may be one that uniquely picks out a particular mode of presentation, or it may simply be the property of being identical to such–and–such mode of presentation. So at one end of the spectrum, the type referred to may constitute a reference to a unique mode of presentation; at the other end, the type might be vacuous, as when the speaker's import in uttering [1] is merely that Ralph believes the proposition that Fido is a dog under some mode of presentation or other. In between, we have reference to substantial types that do not determine unique modes of presentation.

So the representation of [1] as [2] tells us quite a bit. It tells us that a correct compositional truth theory for English must construe 'believes' as a three–place relational predicate holding among believers, Kaplan propositions, and modes of presentation of those propositions. It tells us that [1]'s 'that'–clause, 'that Fido is a dog', is a referential singular term whose referent is the singular proposition $\langle$Fido, doghood$\rangle$, and it tells us that the references of 'Fido' and 'dog' in that 'that'–clause are Fido and doghood, respectively. It tells us that an utterance of the sentence requires reference to a type of mode of presentation, so that an utterance of [1] is true just in case [2], where Φ^* is the mode–of–presentation type referred to in the utterance. And it tells us that this contextually–determined reference to a type of mode of presentation is by a "hidden indexical" in that there is no actual indexical in [1] that carries this reference. In this respect, belief ascriptions are like other sentences containing hidden indexicals. For example, in uttering 'It's raining', the speaker must be referring to some place at which it is raining (typically, this is the speaker's location, but it need not be: a speaker in Chicago may reply 'It's raining' when asked about the weather in New York).

2 Remarks in Support of the Hidden–Indexical Theory

According to the hidden–indexical theory, an utterance of the sentence '*A* believes that *S*' is true just in case the utterer is referring to a type of mode of presentation Φ and the referent of *A* believes the Kaplan proposition referred to by 'that *S*' under some mode of presentation of the type Φ. My claim is that this theory provides the best account of how belief ascriptions must be accommodated in a compositional truth theory for English if English has one. A full-scale defense of this claim would outrun a feasible number of pages, but I would like to give some *sense* of what supports the theory.

One very important thing the hidden-indexical theory tells us is that the 'that'-clause in a paradigmatic belief ascription such as [1] is a *referential singular term.* This at least has the *appearance* of being right: [1] does indeed seem to be telling us one of the things Ralph believes —to wit, that Fido is a dog. On the face of it, [1] says that Ralph stands in the belief relation to the referent of the singular term 'that Fido is a dog'. This appearance that [1]'s 'that'-clause, 'that Fido is a dog', is a referential singular term is sustained by the evident validity of arguments such as:

> Ralph believes that Fido is a dog, and so does Thelma. So, there is something that they both believe (viz., that Fido is a dog).
>
> Ralph believes everything that Thelma says. Thelma says that Fido is a dog. So, Ralph believes that Fido is a dog.
>
> Ralph believes that Fido is a dog. That Fido is a dog is impossible. So, Ralph believes something that is impossible (viz., that Fido is a dog).

For how else are we to account for the validity of these inferences otherwise than on the assumption that 'that'-clauses are referential singular terms? I think that the rhetorical force of this question is correct: the best explanation of the way in which these arguments may be valid does presuppose that 'that'-clauses are referential singular terms.[6]

[6] I say 'the way in which these arguments *may* be valid' because of the alleged hidden-indexical nature of belief ascriptions: qua inference *type*,

It is obvious that we may give simple and straightforward explanations of the validity in question via hypotheses which entail that 'that'–clauses are referential singular terms. The question is whether validity can be accounted for without that assumption. Perhaps, as I myself argued in *Remnants of Meaning*,[7] the arguments are valid only when their quantifications are read *nonobjectually*, say, as something akin to substitutional quantification. This would entail that 'that'–clauses were *not* referential singular terms. Or perhaps, as Graeme Forbes has argued,[8] we can use some very fancy footwork to account for the validity even though [1]'s 'that'–clause disappears on analysis, leaving us with an existential generalization along the lines of

> $(\exists m)(\exists m')$(m is a mode of presentation of Fido & m' is a mode of presentation of doghood & B(Ralph, the proposition that m has m')).

Or perhaps we could salvage enough validity using Davidson's "paratactic" theory of propositional–attitude ascriptions, wherein the only singular term referring to what Ralph believes is the demonstrative 'that', as it occurs in

> Ralph believes that. Fido is a dog.

when its reference is the utterance that follows it of 'Fido is a dog'.

Perhaps; but I doubt it. I have argued elsewhere[9] against Davidson's proposal, and if it were not for present constraints

---

> Ralph believes that Fido is a dog.
>
> So, Ralph believes that Fido is a dog.

is no more valid than

> It's raining.
>
> So, it's raining.

Tokens of these inference types are valid only when premise and conclusion make the same implicit reference.

[7] Cambridge, Mass.: MIT Press, 1987.

[8] "The Indispensability of Sinn", *The Philosophical Review*, XCIX, 4 (1990): 535-563.

[9] *Remnants of Meaning*, ch. 5.

I would offer to explain why neither Forbes' line nor my former self's nonobjectual–quantification line can be made to succeed.[10] But there are the constraints; so I must ask you provisionally to assume that I have justified the assumption that 'that'–clauses are referential singular terms. Then we may move to the question this assumption invites.

If 'that Fido is a dog' is a referential singular term, then we may ask what its reference is. To what, that is, does it refer? This, however, is easy to answer: 'that Fido is a dog' refers to that Fido is a dog; what 'that Fido is a dog' refers to is that Fido is a dog. But what is this thing, *that Fido is a dog*, which is the referent of the 'that'–clause singular term? Well: (i) It is *abstract*, in that it has no spatial location. That Fido is a dog is not under the kitchen sink or anywhere else. (ii) It is *mind– and language–independent* in that it exists in possible worlds in which there are neither thinkers nor speakers. That Fido is a dog is also language independent in that, while it may be the content of a sentence of any language, it itself belongs to no language; it is not a linguistic entity. (iii) It has a *truth condition*, and it has its truth condition *essentially*, in contrast to the contingent way sentences have theirs. It is a contingent fact that 'Fido is a dog' is true iff Fido is a dog; if the conventions of English had been different, it might have had a totally different truth condition or none at all. But it is a necessary truth that that Fido is a dog is true iff Fido is a dog. It follows, then, that the referents of 'that'–clauses

[10] Well, I guess I can afford to mention one especially devastating problem for the Forbes line. Applied to

> Ralph said that Fido is a dog

it yields

> $(\exists m)(\exists m')$(m is a mode of presentation of Fido & m' is a mode of presentation of doghood & S(Ralph, the proposition that m has m')).

The trouble with this is that it is extremely unlikely that Ralph, in his utterance of 'Fido is a dog', will have said any such mode–of–presentation–containing proposition. If he did mean some such proposition, then there would be a specification of what he said that is other than 'that Fido is a dog' and that refers to a mode–of–presentation–containing proposition. But it is clear that there need be no such alternative specification of what he said.

are *propositions*, in the philosophical sense of that term: abstract, mind– and language–independent objects that have essentially the truth conditions they have.

If the hypothesis that 'that'–clauses are referential singular terms is to cohere with the assumption that English has a compositional truth theory, then we must suppose that each 'that'–clause is a semantically complex singular term whose reference is determined by its syntax and the references, or extensions, of its component words. What, we might then ask, are the references of 'Fido' and 'dog' in 'that Fido is a dog', as it occurs in an utterance of [1]?

For all intents and purposes, there are but two possible answers. One, apparently Frege's, is that the terms in a 'that'–clause refer to modes of presentation and the proposition referred to by the entire 'that'–clause is a structured proposition made up of those modes of presentation. With respect to a particular utterance of [1], the Fregean proposal would be that the token of 'Fido' refers to a particular mode of presentation of Fido while the token of 'dog' refers to a particular mode of presentation of doghood, the utterance of [1] thus enjoying the representation

$$B(\text{Ralph}, \langle m_f, m_d \rangle),$$

$\langle m_f, m_d \rangle$ being the mode–of–presentation–containing proposition referred to by the occurrence of 'that Fido is a dog' in the utterance. This will be true, of course, only if m_f and m_d are ways Ralph has of thinking of Fido and doghood respectively.

The merit of this proposal is that it allows us to see how the ancient astronomer was able to believe that Hesperus was Hesperus long before she believed that Hesperus was Phosphorus. Its demerit is that it is false. For consider

[3] Everyone who has ever known her has believed that Madonna was musical.

According to the Fregean proposal, there is a particular mode of presentation m of Madonna and a particular mode of presentation m' of the property of being musical such that the foregoing utterance of [3] is true only if everyone who has ever known Madonna has believed the proposition $\langle m, m' \rangle$. Yet this is surely too strong a requirement on the truth of [3]. It requires that everyone who has ever known Madonna shared a single way of thinking of her and a single way of thinking of the property of being musical, and this is most unlikely given that there may have been people who knew her as a child and then died and that someone like Helen Keller may have been among them. If the Fregean proposal were correct, we could prove that [3] was false just on the basis of the virtual certainty that, whatever modes of presentation turn out to be, there will not be one single mode of presentation of Madonna shared by all who ever knew her. But it seems that we cannot disprove [3] in this way, and this because we would count it as true if each person who has ever known Madonna thought of her as being musical *under some mode of presentation or other.*[11]

Our question is: What are the references of 'Fido' and 'dog' in [1]'s 'that'–clause on the assumption that it refers to a proposition that is true just in case Fido is a dog? The second possible answer is that they are Fido and doghood, respectively, ⟨Fido, doghood⟩ being the referent of 'that Fido is a dog'.[12] Certainly this offers a proposition with the right truth conditions. The prima facie problem with this proposal

---

[11] Here is a longer way of showing that 'that'–clauses do not refer to mode–of–presentation–containing propositions. Intuitively, 'believes that Fido is a dog' may be univocally true of people who think of Fido and doghood in radically different ways (you and Helen Keller may be among them) and who do not share any single mode of presentation for either Fido or doghood. This shows that 'that Fido is a dog' makes no context–independent reference to a mode–of–presentation–containing proposition. At the same time, a speaker may truly say that so–and–so believes that Fido is a dog even though she is not in a position to refer to any particular mode of presentation so–and–so has for either Fido or doghood. This shows that the 'that'–clause makes no context-dependent reference to a mode-of-presentation-containing proposition. And if it makes neither a context–independent nor a context–dependent reference to such a proposition, then it makes no reference to one.

[12] Friends of possible–worlds semantics or of counterpart theory will think my rush to singular propositions a bit cavalier. Why not say that the reference of 'Fido' is a function from possible worlds to referents of 'Fido' in those worlds —Fido himself, perhaps, in each world, or variable counterparts of our Fido? First, I think importation of possible worlds

is that it is apt to suggest —and has been taken by many to suggest— that the logical form of [1] may be represented as

[4] $B(\text{Ralph}, \langle\text{Fido}, \text{doghood}\rangle)$.

And you may be very quickly reminded of the well-known problem with this representation if I momentarily change the example. Surely, Lois Lane does not believe that Clark Kent flies, though she does believe that Superman flies. But this is impossible according to the present proposal, since, as Clark Kent = Superman, the proposition that Superman flies = the proposition that Clark Kent flies. Likewise, reverting now to the ongoing example, Ralph may believe that Fido is a dog without believing that Fido is a shmog, and he may believe that Fido is male without believing that Fi Fi is male. Yet all this is impossible according to the theory in question, since it implies that the proposition that Fido is a dog = the proposition that Fido is a shmog and that the proposition that Fido is male = the proposition that Fi Fi is male. The problem is even manifested in belief ascriptions involving indexical reference. For example, pointing to one mug shot, Thelma might say, 'I believe that he is the culprit', while pointing to another, she might say, 'I neither believe nor disbelieve that this guy is the culprit; I simply can't tell'. Intuitively, both of her utterances are true, even though, as it may happen, the same man is referred to in each utterance; but this, again, is incompatible with the proposal in question.

Nathan Salmon has tried to defend the extreme 'Fido'-Fido theory of belief that would represent [1] as [4] by trying to explain away the patina of counterintuitiveness adhering to the theory's consequences.[13] But I think he has not succeeded, and in "The 'Fido'-Fido Theory of Belief" I try to

---

is unmotivated, though I cannot argue that here. Second, I think any prima facie plausible theory that used possible worlds and propositions that were constructions based on them would, when the dust settled, simply yield alternative styles of bookkeeping. Third, nothing that I really care about changes if, say, you opt for a version of the hidden-indexical theory wherein 'Fido' in a 'that'-clause refers to that constant function that maps each possible world onto our Fido (but the counterpart theory might be a version of the rejected proposal that the referent is a mode of presentation of Fido).

[13] *Frege's Puzzle* (Cambridge, Mass.: MIT Press, 1986).

show why.[14] We really do need a theory that allows us to see how an utterance of

> Lois believes that Superman flies but doesn't believe that Clark Kent flies

can be literally true.

Here, then, is our situation. We are motivated to see [1]'s 'that'–clause as a referential singular term whose reference is the proposition that Fido is a dog. Since we are assuming that English has a compositional truth theory, we must hold the reference of this 'that'–clause to be determined by its syntax and the semantic values the words in the 'that'–clause have in it. Those semantic values —those references— cannot be modes of presentation, and the only viable option is that they are Fido, for 'Fido', and doghood, for 'dog'. At the same time, we cannot accept the representation of [1] as [4], for we also want to allow that Ralph does not believe that Fido is a shmog.

Enter now on its white charger the hidden–indexical theory, which satisfies all desiderata. This theory, you will recall, represents the logical form of an utterance of [1] as [2], where Φ^* is an implicitly referred to and contextually–determined type of mode of presentation. Thus, 'that Fido is a dog' in [1] is a referential singular term whose referent is the singular proposition ⟨Fido, doghood⟩, and thus the referents of 'Fido' and 'dog' in that 'that'–clause are Fido and doghood, respectively. At the same time, we can see how an utterance of

[5] Ralph does not believe that Fido is a shmog

can also be true, even though, as shmoghood = doghood, 'shmog' in [5] has the same reference as 'dog' in [1], so that the 'that'–clause in each sentence refers to one and the same singular proposition, to wit, ⟨Fido, doghood (i.e., shmoghood)⟩. That the utterances of [1] and [5] can both be true is due, of course, to the hidden–indexical feature: in the normal case, the type of mode of presentation referred to in an

---

[14] See also Salmon's reply, "Illogical Belief," *Philosophical Perspectives*, III (1989): 243-285.

utterance of [1] will be different from that referred to in an utterance of [5]. Likewise in the other examples discussed, and likewise as regards the London/Londres and Paderewski examples that puzzled Kripke.[15] These are no puzzles for the hidden–indexical theory.

3 Three Problems for the Hidden–Indexical Theory

So much for what recommends the hidden–indexical theory. I shall now raise three problems for the theory.

3.1 THE CANDIDATE PROBLEM

This problem arises for any theory that appeals to modes of presentation. It is a problem I have discussed at length elsewhere,[16] and I will say only enough about it now to indicate its nature. In a nutshell, the candidate problem is that there is some question whether there is a plausible candidate for what modes of presentation are.

The notion of a mode of presentation is, we saw, a functionally defined technical notion: a mode of presentation is whatever plays the mode–of–presentation role defined by Frege's Constraint. So we may ask what things play that role. This is like asking what genes are, the notion of a gene being functionally defined as whatever is responsible for the transmission of hereditary characteristics. There the answer turned out to be segments of DNA molecules. What, though, is the answer to the mode–of–presentation question? There is no dearth of proposed answers:

Modes of presentation are *individual concepts*, uniqueness properties of the form *the property of being the unique instantiator of such–and–such property.*

---

[15] "A Puzzle about Belief", in A. Margalit, ed., *Meaning and Use* (Dordrecht: Reidel, 1979).

[16] "The Basis of Reference," *Erkenntnis*, XIII, 1 (1978): 171-206; *Remnants of Meaning*, esp. ch. 3; and "The Mode–of–Presentation Problem".

Modes of presentation are simply *general properties* that make no pretense of applying uniquely to the things they are modes of presentation of.

Modes of presentation are *percept tokens*, actual instances of sensory experience.

Modes of presentation are "*stereotypes*", perhaps what cognitive psychologists call *prototypes*.

Modes of presentation are "*characters*", functions from contexts of the utterance of an expression to the expression's contents in those contexts.

Modes of presentation are *public language expressions*.

Modes of presentation are *Mentalese expressions*, formulae in the neural language of thought.

Modes of presentation are *functional roles*, perhaps conceptual roles of Mentalese expressions.

Modes of presentation are *causal chains* linking Mentalese names and predicates to the objects and properties they stand for.

And finally there is the *no–theory theory of modes of presentation*, which rejects what I have elsewhere called the *Intrinsic–Description Constraint*.[17] This constraint holds that if a thing is a mode of presentation —if, that is, it plays the mode–of–presentation role— then it must be intrinsically identifiable in a way that does not describe it as a mode of presentation or as a possible mode of presentation. If a thing is a mode of presentation, then it must be intrinsically identifiable as some other kind of thing. Rejecting this constraint, the no–theory theorist of modes of presentation holds that modes of presentation are modes of presentation and there is an end to it. You simply cannot say what they are in any other terms.

But I think the Intrinsic–Description Constraint *is* well motivated, and I think there are compelling objections against each of the foregoing candidates but one. The exception is the view that modes of presentation are functional roles, perhaps conceptual roles of Mentalese expressions. This view of what modes of presentation are has intrinsic

[17] "The Mode–of–Presentation Problem".

problems and it does not cohere with every theory of the logical form of belief ascription; but its problems do not *refute* it and it does cohere with the hidden–indexical theory.[18]

In any case, the hidden–indexical theory entails that believing is a relation among a believer, a proposition, and a mode of presentation of that proposition, which mode of presentation is in turn determined by modes of presentation of the proposition's components. It is hard to see how we can reasonably accept this theory without some well–motivated idea of what these modes of presentation are supposed to be.

3.2 THE MEANING–INTENTION PROBLEM

This problem is that one may reasonably doubt that belief ascribers mean what the hidden–indexical theory requires them to mean when they ascribe beliefs. Let me explain.

Floyd's sister calls him in Chicago and asks about the weather. Floyd replies, "It's raining". Here Floyd refers to Chicago, and this by virtue of the fact that in uttering 'It's raining', Floyd *means* that it is raining in Chicago. In other words, Floyd counts as having referred to Chicago because the proposition he meant is about Chicago. Notice that there is no difficulty whatever in ascribing to Floyd the

---

[18] This corrects a flawed argument in "The Mode–of–Presentation Problem" which was intended to show that modes of presentation cannot be conceptual roles of Mentalese expressions. The argument overlooked what must be the intended generalization of the hidden-indexical theory for belief ascriptions with embedded 'that'–clauses —namely, that a belief ascription with n 'that'–clauses embedded in the primary 'that'–clause must be making $n+1$ implicit references to types of modes of presentation. Thus, the logical form of an utterance of

Floyd believes that Lois believes that Superman flies

is

$$(\exists m)(\Phi m \ \& \ B(\text{Floyd}, [(\exists m')(\Phi' m' \ \& \ B(\text{Lois}, [\text{Superman flies}], m'))], m)),$$

where Φ and Φ' are implicitly referred to and contextually–determined types of modes of presentation (and where '[...]' is a singular–term forming operator with the force of 'the proposition that ...'). If this inevitable complexity weakens the plausibility of the hidden–indexical theory, then so be it.

propositional speech act in question: he clearly intended his sister to believe that it was raining *in Chicago*, and he is quite prepared to tell you that this is what he meant, what he implicitly said, and what he intended her to be informed of. Because these things are so clear, Floyd's utterance is a paradigm of implicit reference.

Now for a different case. During a casual conversation about air–fare bargains, Flora says,

[6] Harold believes that TWA is offering a New York–Paris return fare for $318.

According to the hidden–indexical theory, there is a property Φ of modes of presentation of the proposition that TWA is offering a New York–Paris return fare for $318 such that Flora referred to Φ in her utterance. If this is true, then, presumably, it is because Flora, in producing her utterance, meant some proposition about Φ in just the way that Floyd, in his utterance, meant some proposition about Chicago. But, we shall see, it is doubtful that Flora meant any such thing.

We had no trouble saying what Floyd meant in uttering 'It's raining': he meant that it was raining in Chicago. But what that implies a reference to a type of mode of presentation does Flora mean in uttering [6]? It must surely be

[7] that there is something that both has Φ and is such that Harold believes the proposition *that TWA is offering a New York–Paris return fare for $318* under it,

for some particular property Φ of modes of presentation of the proposition that TWA is offering a New York–Paris return fare for $318. But —and this is where the trouble begins— neither Flora nor her audience is aware of her meaning any such thing. Floyd is consciously aware of both the *form* of the proposition he meant (he is aware that he meant that it is raining *in* Chicago) and the implicitly referred to thing the proposition is about (he is aware that he meant that it is raining in *Chicago*). But it is doubtful that the nonphilosopher Flora has conscious access to the form of [7], and it is especially doubtful that she has conscious awareness of referring to any mode–of–presentation property. That is to say, if she did mean a proposition of form [7], then she has no conscious awareness of what property Φ that proposition is about, and no conscious awareness, therefore, of what she meant in uttering [6].

Neither we nor Flora can say what mode–of–presentation property was referred to in her utterance of [6]. Just try to say it. The referred–to property, if it exists, would be given by a completion of the form

> the property of being an m such that to believe the proposition *that TWA is offering a New York–Paris return fare for $318* under m requires thinking of TWA, the offering relation, New York, Paris, the US dollar currency, the number 318 [and so on for the other components of the proposition] in ... ways, respectively.

But the nonphilosopher Flora has no access even to this *form* of specification, and who among us can offer to replace the three dots? Now of course it is in principle possible to have knowledge of such a property even though one cannot specify it in the forgoing way. One might, for example, have knowledge of it under some quite extrinsic description. But I submit that it will be obvious on reflection that Flora has no such alternative way of explicitly picking out a property of modes of presentation that she is implicitly referring to in her utterance of [6].[19] There is no sentence σ such that Flora can say 'I meant that σ', where 'that σ' explicitly refers to a proposition of [7]'s form.

Thus, if the hidden–indexical theory is correct, then Flora has no conscious awareness of what she means, or of what she is saying, in uttering [6], and this is a *prima facie* reason to deny that she means what the theory is committed to saying she means.

---

[19] Besides, it is quite against the spirit of the hidden–indexical theory that the type of mode of presentation referred to should be one that the speaker knows only under some nonintrinsic mode of presentation. The idea of the theory is that, though one might not know the particular mode of presentation under which so–and–so believes that such–and–such, still one has access to some interesting *description* that the mode of presentation satisfies.

There is more than one way a hidden–indexical theorist might respond to this *prima facie* problem, but let me go directly to what I think is the theorist's best bet. This response is to concede that Flora is not *aware* of meaning a proposition of form [7] under any mode of presentation but to insist that she nevertheless meant such a proposition. The idea is to appeal to socalled "tacit belief" and "tacit intention". This idea has its origin, I believe, in Chomsky's controversial claim that speakers know a grammar of their language. To this it was protested that in no ordinary sense of 'know' do speakers know any such thing. When the dust settled, many theorists supposed that the best construal of Chomsky's claim was that the correct theory of language processing implies that a speaker has an internal representation of the grammar of her language, where this internal representation is stored not as a *belief*, but as what Stich was later to call a *subdoxastic state*.[20] Subdoxastic states are representational states that are belieflike in some respects but differ from paradigmatic beliefs in their unavailability to consciousness and the way they are inferentially isolated from beliefs (for example, if you believe that if *p*, then *q*, and believe *p*, you are likely to infer *q*; but you will not make this inference if the state representing *p* is subdoxastic). Now the notion of a tacit propositional attitude can admit of a spectrum of cases, from the Chomsky–required subdoxastic states at one end, to, at the other end, states that may reasonably be claimed to be subsumable under our current propositional–attitude concepts. What is evidently unifying about the notion of a tacit propositional attitude is that a tacit propositional attitude is unavailable to consciousness and that one cannot be made irrational by conflicts between those of one's thoughts that are available to consciousness and those that are merely tacit. At any rate, the present proposal is that it is thanks to her *tacit* beliefs and intentions in uttering [6] that Flora means a proposition of [7]'s form even though she is not aware of meaning any such thing.

---

[20] "Beliefs and Subdoxastic States", *Philosophy of Science*, XLV (1978): 499-518.

One trouble with the tacit–intention proposal is that it induces a rather radical *error theory*: not only do ordinary belief ascribers have no conscious knowledge of what they are asserting, they also turn out not to have the conscious thoughts they think they have. Flora clearly *thinks* she has conscious knowledge of what she is asserting in uttering [6]. She is quite prepared to say, "Look, what I am saying, and all that I am saying, is that Harold believes that TWA is offering a New York–Paris return fare for $318". In other words, she thinks she is consciously aware of what she is asserting in uttering [6], but the tacit–intention line implies that here she is in error: the only proposition she asserts in uttering [6] —viz., some proposition of form [7]— is not anything she is conscious of. What makes this error aspect of the tacit–intention proposal problematic is not merely that it riddles the propositional–attitude ascriptions of ordinary speakers with error; it also forces us to qualify our views about first–person authority in an important way. Flora does not have the privileged access to what she consciously means, intends, and believes in uttering [6] which one might reasonably have supposed to be part of a normal person's functional architecture.

A more serious trouble with the tacit–intention proposal actually constitutes a direct objection to the claim that speakers have the meaning intentions the hidden–indexical theory requires them to have, whether or not those intentions are claimed to be tacit. But the objection is best presented here, in response to the tacit–intention proposal, since it is obvious that speakers do not consciously mean what the hidden–indexical theory entails that they mean. The objection takes hold even if it is conceded that ordinary people do have some sort of indirect awareness of modes of presentation, which is all to the good, as it is not out of the question that one could motivate the idea that Flora tacitly *believes* that Harold believes the proposition about TWA under a certain type of mode of presentation. The point to be pressed, however, is that considerably more is required of tacit *meaning*. I shall explain.

Meaning entails audience–directed intentions, and one cannot mean something without intending to be understood.

Part of meaning that such and such is intending one's audience to recognize that that is what one meant, and —a corollary— part of referring to a thing is intending one's audience to recognize that reference. Consequently, the present tacit–intention proposal requires there to be some type of mode of presentation Φ such that Flora tacitly means that Harold believes that TWA ... under a mode of presentation of type Φ, and tacitly intends her audience to recognize (no doubt tacitly) that she means that Harold believes that TWA ... under a mode of presentation of type Φ. But as Jim Higgenbotham pointed out to me, this puts the implausibility of Flora's *meaning* a proposition of form [7] on a par with that of a certain way of trying to extend the description theory to "incomplete" definite descriptions. The idea there is that when a speaker utters a sentence like 'The dog has fleas', then there must be some property Φ such that she means the proposition that the thing that is uniquely a dog and Φ has fleas. What makes this so implausible is that there will typically be a number of potentially completing descriptions that are equally salient in the context (e.g, *the dog that I own*, *the dog we are both looking at*, *the spotted dog in the green chair*, and so on). Consequently, it is highly implausible that there will be one such description such that the speaker intends it to be understood between her and her audience that she means a proposition containing that very description.[21]

Now just the same problem arises with the meaning claim required by the hidden–indexical theory. For assume that propositions are believed under modes of presentation. If a proposition is believed under one mode of presentation, then it will typically be believed under many modes of presentation. Further, each of those modes of presentation will instantiate infinitely many *types* of modes of presentation, many of which will be equally salient in the communicative context. This makes it extremely implausible that of all the equally salient types of ways that Harold has of believing the proposition about TWA, Flora should mean —and intend to

---

[21] See my "Indexicals and the Theory of Reference", *Synthese*, XLIX, 1 (1981):43-100; esp. pp. 77-8.

be taken to mean— a proposition about one definite one of them.

A little thought experiment should make the objection vivid. When considering the hypothesis that a certain expression has a certain semantic property, it is often helpful to introduce an expression that has that semantic property by stipulation. Presently in debate is the hidden–indexical theory's claim that one uttering [6] would mean a proposition of the form [7] displays. So let us introduce 'shmelieves' as a term that satisfies the hidden–indexical theory by stipulation: to utter a sentence of the form

> *A* shmelieves that *S*

with its literal meaning requires a speaker to mean

> that there is something that both has Φ and is such that *A* believes the proposition that *S* under it,

for some particular contextually determinate property Φ of modes of presentation of the proposition that *S*. The hidden–indexical theory may or may not be true of '*A believes* that *S*', but it is true by stipulation of '*A shmelieves* that *S*'.

So far so good, but let me now try to use this newly minted coin to make an assertion:

> Placido Domingo shmelieves that Luciano Pavarotti will be arriving at Orly airport today at around 4 PM Paris time.

Very well, what assertion did I make? What is the type of mode of presentation Φ such that I meant that Placido Domingo believes the proposition *that Luciano Pavarotti will be arriving at Orly airport today at around 4 PM Paris time* under a mode of presentation that has Φ? Remember, Φ is made up of types of modes of presentation for each of the constituents of the proposition my 'that'–clause refers to: Luciano Pavarotti, the arrival relation, Orly airport, the number 4, Paris, etc. The question, of course, cannot be answered. For although 'shmelieves' was stipulatively given a perfectly good hidden–indexical semantics, I was not, in uttering the displayed sentence, in a position assertively to use it, and this because I was not in a position to refer to a contextually–determinate type of mode of presentation. None is sufficiently salient to enable you, my audience, to identify it as the one I meant, and this notwithstanding the fact that we understand all the concepts involved and everything has been painstakingly raised for us to the level of conscious awareness.

The application to the hidden–indexical theory should be clear. Although I cannot utter the forgoing 'shmelieves' sentence and mean what its literal meaning requires me to mean, there are no such impediments to my uttering 'Placido Domingo *believes* that Luciano Pavarotti will be arriving at Orly airport today at around 4 PM Paris time'. I can perfectly well utter this and hope to have my literal assertion be perfectly well understood by you. Since 'shmelieves' is by stipulation what 'believes' would be if the hidden–indexical theory were true, this appears to show that the theory is not true. The best account of why neither Flora nor her audience (who understood her perfectly well) is aware of her meaning a proposition of form [7], as the hidden–indexical theory requires her to mean, is not that the requisite meaning intentions have gone tacit; it is that she does not mean any such thing.

3.3 THE LOGICAL–FORM PROBLEM

According to the hidden–indexical theory, 'believes' is a three–place relational predicate. But this might strike one as fishy for the following reason. If 'believes' really were three–place, then it ought to have occurrences where its three–place form is explicit, as the three–placedness of 'gives' is made explicit when we move from 'She gave the house' to 'She gave the house to her husband'. Yet in what analogous way can we make explicit the three–placedness of 'Ralph believes that Fido is a dog'? Perhaps in the way of

[8] Ralph believes that Fido is a dog *in way w/under mode of presentation m*.

There are, however, two problems with this.

First, this is no ordinary language specification but technical jargon. If one is told that Mary gave the house, one can ask, "To whom did she give the house?" But what is the corresponding ordinary language question for 'Ralph believes that Fido is a dog'? Ask a nonphilosopher, "In what way, or under what mode of presentation, does Ralph believe this?", and your best answer will be a puzzled look.

Second, [8] does not look like a specification of a three–place relation. It looks more like a two–place relation with *an adverbial qualifier*. In other words, it is to be assimilated to

Ralph kissed her in the most exciting way,

where 'the most exciting way' is merely part of an adverb, rather than to

[9] Mary gave the house to her husband,

where 'her husband' is an argument of the verb 'to give'. Evidence for this assimilation comes from well–known constraints in syntactic theory which distinguish between arguments and adverbs in their ability to extract from 'whether'–clauses.[22] Thus, the argument status of 'her husband' in [9] is revealed in the fact that we can answer 'Her husband' in response to the question

To whom did you wonder whether Mary gave the house?

And the non–argument, adverbial status of 'way w/mode of presentation m' in is revealed by the fact that it cannot be given in answer to the question

In what way/under what mode of presentation did you wonder whether Ralph believes that Fido is a dog?

4 The Importance of the Issue

I have proposed that if natural languages have compositional truth theories, then the hidden–indexical theory is the correct

[22] See N. Chomsky, *Barriers* (Cambridge, Mass.: MIT Press, 1986). I owe this reference and the point about 'whether'–clauses to Bob Fiengo.

account of the accommodation of belief ascriptions in them. I have also offered some reason for thinking the hidden-indexical theory is false. If it is both conditionally correct and false, then it follows that natural languages do not have compositional truth theories. This is one reason the issue of the semantics of belief ascriptions is important. There is, however, another reason for the importance of the question of the accommodation of belief ascriptions in a compositional truth theory; it has to do with the bearing of this question on the vexing question about the form that a compositional *meaning* theory for a language must take. I shall explain.

It is commonly assumed that each natural language L has a *compositional meaning theory*, this being a finitely axiomatizable theory of L whose theorems ascribe to each sentence of L its meaning in L. Those who make the assumption reason that without it there will be no explaining either the platitude that the meaning of a sentence is determined by its syntax and the meanings of its component words or the fact that we have the ability to understand indefinitely many novel sentences. But while it is generally agreed that every natural language enjoys a compositional meaning theory, no three philosophers can be made to agree on the form such a theory must take. For a start, it is not even clear what form the *theorems* of a meaning theory would have to take. You might at first blush suppose that a compositional meaning theory for L would generate theorems of the form

'S' means in L that such and such,

but this could hardly be the form of a theorem for sentences like 'He was there' or 'Stop singing, you idiot!', and it is anything but clear that the word 'means' should ever be used in saying what a sentence means. After all, to know the meaning of a sentence is to be able to know what a speaker is saying in uttering the sentence; so it is reasonable to suppose that a compositional meaning theory would have to ascribe to each sentence a feature knowledge of which would enable us to know what would be said in the utterance of any sentence having that feature. But there is no agreement as to what such a feature would be like, or that the word 'meaning' would need to be used to ascribe it.

Those who assume that each natural language has a compositional *meaning* theory typically assume that each natural language also has a compositional *truth* theory, in the sense already glossed. After all, it is part of the *meaning* of 'Snow is white' that an utterance of it is true iff snow is white, so if it is reasonable to think a compositional theory can assign meanings to sentences, then it is reasonable to think one can assign truth conditions to them, too. What we are now in a position to see is that if we know how to accommodate belief ascriptions in a compositional truth theory, then we will be able to answer the question about the form that a compositional meaning theory must take. This ought at first to seem a little surprising. Merely knowing how to accommodate 'Fido is a dog' in a truth theory would not per se tell us the first thing about how to accommodate that sentence in a *meaning* theory. So how can knowing how to accommodate *belief ascriptions* in a truth theory tell us how to accommodate *any* sentence in a meaning theory?

I shall forgo a general answer, but we can see straightway how the hidden–indexical theory of belief ascription constrains the nature of a meaning theory. For what applies to belief ascriptions applies, mutatis mutandis, to all propositional–attitude ascriptions involving 'that'–clauses. What holds for 'Ralph believes that Fido is a dog' also holds for 'Ralph says that Fido is a dog'. The theory tells us that the 'that'–clause in a saying ascription refers to a Kaplan proposition. This means that to know what a speaker said in the utterance of a sentence is to know what Kaplan proposition she said. Since to understand a language is to have the ability to understand utterances in it, and since to understand an utterance is to know what the speaker said in it, we can see that, if the hidden–indexical theory is correct, then the main task of a compositional meaning theory for a language will be the construction of a finitely specifiable correlation that relates each sentence of the language to what Kaplan has called a *character* —a function that, in effect, maps each utterance of the sentence onto the Kaplan proposition that is the propositional object of the primary speech act performed in the utterance.[23] In other words, if the hidden–indexical theory of the accommodation of propositional–attitude ascriptions in a truth theory is correct, then so is the "direct reference" approach to a compositional meaning theory, and in this way we can see how the hidden–indexical theory constrains the nature of a meaning theory.

It may be worth noting that this point I have been laboring reverses the usual procedure of those who argue for the direct reference treatment of 'that'–clauses implied by the hidden–indexical theory. That procedure begins by assuming a direct reference approach to meaning and then draws the consequence as regards the propositional content of propositional–attitude ascriptions, which consequence entails that 'that'–clauses are referential singular terms whose referents are Kaplan propositions. My procedure has been different, and I think less contentious. I started with no assumptions about "content" or meaning. My question was wholly about the *references* that must be assigned to words in a belief ascription on the assumption that languages had compositional *truth* theories, theories whose theorems are of the extensional paradigm

'La neige est blanche' is true in French iff snow is white.

It was then argued that, relative to the assumption, 'that'–clauses had to be construed as referential singular terms whose *referents* are Kaplan propositions. Then, in the final step, this premise about *reference* was used to derive the general point about *meaning*.

5 A Paradox of Meaning

If the hidden–indexical theory is correct, then we not only know how to accommodate belief ascriptions in a compositional truth theory for English, we also know the form that a compositional meaning theory for English must take. But

---

[23] David Kaplan, "Dthat", in P. Cole, ed., *Syntax and Semantics 9: Pragmatics* (New York: Academic Press, 1978).

I did not argue that the hidden–indexical theory is correct. I merely argued that it is the correct theory relative to the assumption that natural languages have compositional truth theories, and I proceeded to raise problems for the theory. Evidently, we have the makings of a paradox, a set of mutually inconsistent propositions each of which enjoys some plausibility. Specifically, we have this paradox set:

(1) 'That'–clauses refer to propositions.

(2) The reference of a 'that'–clause is determined by its syntax and the references its words have in it.

(3) There is no tenable account of the compositional determination of 'that'–clause reference consistent with (1) and (2).

We know the prima facie case for (1): 'that'–clauses behave syntactically like singular terms, and their being referential singular terms is arguably needed to account for the validity of inferences like the ones considered in section 2.[24] We know, too, the prima facie case for (2): given that 'that'–clauses are referential singular terms, and given the obvious fact that they are not semantically *simple* singular terms, (2) is required by the assumption that natural languages have compositional truth theories. And the prima facie case for (3) has been the main business of this paper. For suppose we accept (1) and (2). Then we must find references for the terms in a 'that'–clause. We have seen that those references cannot be modes of presentation, and the remaining option is that they are constituents of Kaplan propositions —Fido and doghood, in the case of 'that Fido is a dog'. But this remaining option is plausible, I have argued, only when wrapped in the hidden–indexical theory, and that theory, I have also argued, is in the end not plausible.

---

[24] Besides, comparable paradoxes, derivable from the materials already used, can be constructed given familiar ways of trying to deny (1). This is because any plausible account of the semantics of belief ascriptions that comports with compositional semantics will need to claim an indexical reference to a mode of presenta-tion or type of mode of presentation.

To solve our paradox is to reject one of (1)–(3) as the odd man out. Well, which one? One possible solution is to reject (2) while keeping (1) and (3). This solution would have at least two attractive features. First, it allows one to preserve the thesis that a sentence of the form '*A* believes that *S*' has the structure it appears at face value to have —a two–place verb linking two arguments. And second, it allows one to say that the proposition that Fido is a dog the proposition that Fi Fi is a shmog without having to find four different things to be the semantic values of 'Fido', 'Fi Fi', 'dog' and 'shmog' in those 'that'–clauses.[25]

Nevertheless, most philosophers would deem this the least likely solution to the paradox. In the first place, it directly requires denying that natural languages have compositional truth theories, and indirectly requires denying that they have compositional meaning theories.[26] But the assumption that they have them is apt to seem eminently well motivated. For example, without a compositional meaning theory, how can we hope to explain the *systematicity* of language, the fact that, roughly speaking, the meaning of a sentence is determined by features of its words and structure, features that play the same meaning–determining roles in every sentence in which a word or structure having them occurs? In the second place, accepting that 'that'–clauses refer to propositions while denying that the reference of a 'that'–clause was determined by its syntax and the references of its parts would require a conception of propositions as unstructured, non–compositionally–determined entities. On the traditional conception, propositions are both structured (they have components arranged in a certain order) and compositionally determined in the sense that there is a finitely speci-

[25] See the examples in section I.

[26] Rejecting (2) (while accepting (1)) directly requires denying that languages have compositional truth theories, for if languages had such theories, then the reference of a 'that'–clause would have to be determined by its syntax and the references of its parts. It indirectly requires denying compositional meaning theories by this contraposition: since meaning determines truth conditions, if a language has a compositional determination of meaning, then it also has a compositional determination of truth conditions.

fiable function from sequences of propositional determinants —the referents of words in 'that'-clauses, if the references of 'that'-clauses are compositionally determined in the standard truth-theoretic way —onto the infinitely many propositions our 'that'-clauses can refer to. But the idea that the reference of a 'that'-clause was both to a proposition and not compositionally determined could be sustained only if the propositions referred to by 'that'-clauses were themselves undetermined by any finitely specifiable function from things that could be the references of words in 'that'-clauses. Yet this conception of propositions as, so to say, structureless blobs would seem to require some stronger motivation than its ability to occupy a position in the logical space of possible solutions to our paradox. And it must answer some challenging questions, such as:

If the reference of a 'that'-clause is not determined by its syntax and the references of its parts, then how *is* it determined? What makes the proposition that Superman flies the referent of 'that Superman flies'?

We know how to define truth for the familiar sorts of propositions that comport with compositional semantics, but how are we to define truth for these new creatures?

What individuates propositions? What makes the proposition that Superman flies distinct from the proposition that Clark Kent flies?

Despite all this, I am inclined to think that the solution that keeps (1) while rejecting (2) because of (3) is one that deserves to be taken seriously. I say this because *I* do not know how to answer the arguments for (1) and (3), but I think I may know how to reply to the two problems with rejecting (2). In fact, I see the answer to both those problems as hanging essentially together.

Elsewhere[27] I have argued that Mentalese, one's inner system of mental representation, must have a compositional *supervenience* theory, whether or not it also has a composi-

[27] *Remnants of Meaning*; "Does Mentalese Have a Compositional Semantics?", in B. Loewer and G. Rey, eds., *Meaning in Mind: Essays on the Work of Jerry Fodor* (Oxford: Blackwells, 1991); and "Actual–Language Relations", forthcoming in *Philosophical Perspectives*, 7, 1993.

tional meaning theory, and that this will explain the systematicity (etc.) of Mentalese. And I have further argued that the systematicity (etc.) of one's public language can be explained in terms of the systematicity (etc.) of Mentalese. Roughly speaking, a compositional supervenience theory for Mentalese is a finitely axiomatizable theory of Mentalese whose axioms assign physical properties to the words and primitive structures of Mentalese and whose theorems assign to each Mentalese sentence not its meaning, but rather the physical property on which its meaning supervenes. Such a theory will not be a compositional meaning theory, since it will ascribe no semantic properties, but it will explain the systematicity of Mentalese in the straightforward sense that it will be metaphysically impossible for the theory to be true and for Mentalese not to be systematic.

Although a compositional supervenience theory is not per se a compositional meaning theory, the question remains whether the former determines the latter —whether, that is, it is possible for Mentalese to have a compositional supervenience theory without also having a compositional meaning theory. The answer is that this depends entirely on the nature of propositions. If propositions are themselves compositionally–determined entities, then it will be possible to pair off propositional determinants —which on most views will be constituents of propositions —with the physical properties assigned to words in a compositional supervenience theory in such a way as to determine a compositional meaning theory. But if propositions are not themselves compositionally–determined entities, then one will have a compositional supervenience theory without also having a compositional meaning theory, and the compositional supervenience theory will continue to explain those features of languages which have seemed to call for a compositional meaning theory.

So the issue of compositional semantics turns entirely on the nature of propositions. This means that there is no good argument from the need for compositional semantics to the need for compositionally–determined propositions, since languages will need compositional semantics only if propositions are compositionally determined. Thus one should be unimpressed that standard views of propositions see them as compositionally determined, for those views have devised their conception of propositions so as to comport with compositional semantics. The real question, and in the end the only question, concerns the status of *non*–compositionally–determined propositions apart from the issue of compositional semantics. Can this conception be seen to enjoy a positive motivation, and can the questions that appear to challenge it be answered? I am hopeful that the answer is yes. The positive conception of propositions I envisage is a deflationary conception of their ontological status as being so thinly based as not to support any metaphysical view of their nature or individuation that is not determined by the linguistic move that allows us pleonastically to move back and forth between the sentence '*S*' and the singular–term–introducing sentence 'It is true that *S*'. For this in itself sustains no compositionally–determined conception of the referent of 'that *S*'. But I am painfully aware that what is most apt to seem thin is what I am presently saying in defense of these thinly conceived propositions, especially as there is no very brief way to offer what I take to be the answers to such challenging questions as those already mentioned. All that I have done in this paper is to generate, through the travails of the hidden–indexical theory, a paradox of meaning. Its solution, and especially the one I favor, remains, as they say, a research program in progress. But whatever one thinks of this final section's bold remarks, I at least hope it will be agreed that the hidden–indexical theory and its obstacles present a challenge that cannot be ignored.[28]

---

[28]I am much indebted to the participants in the seminar I gave on these topics in 1991 at the CUNY Graduate Center. Various versions of this paper were read at the University of Wisconsin at Madison, the University of Chicago, the 1991 SOFIA conference in Salamanca, Spain, the University of Connecticut, the Autonomous National University of Mexico, and Rutgers University. I benefitted considerably from the discussions at these talks, and especially from the comments of Ned Block, James Higgenbotham, Richard Jeffrey, Jaegwon Kim, Ernest LePore, Richard Mendelsohn, and Ernest Sosa. Virtually all of this article up to the concluding section was previously published as "Belief Ascription" in *The Journal of Philosophy*, Vol. LVIII, 10 (October 1992), 499-521.

PHILOSOPHICAL ISSUES, 3
Science and Knowledge, 1993

Scientific Conceptions of Language and Their Philosophical Import

Paul Horwich

The purpose of this paper is to explore the apparent conflict between Quine's denunciation of the analytic–synthetic distinction and the potentially justified postulation of that distinction in cognitive science.[1] I began thinking about this topic after hearing a lecture on it by Noam Chomsky.[2] He pointed out the prevalent use of the analytic–synthetic distinction in linguistics; he noted that under Quine's influence philosophy has pretty much abandoned it; and he drew the

[1] For a complementary discussion of this topic see "Chomsky versus Quine on the Analytic–Synthetic Distinction", *Proceedings of the Aristotelean Society*, 1991. That was the paper I gave in Salamanca and on which Paul Boghossian commented. It could not be published in these proceedings. However the present piece draws the same conclusions in almost the same way, and differs only in emphasis from the Aristotelian Society paper.

[2] This lecture took place at MIT in the Spring of 1991. Chomsky has responded in print to Quine's critique of meaning in various places. See for example *Rules and Representations* (New York: Columbia University Press, 1978).

conclusion that either philosophy or linguistics must be badly off track. It's not hard to guess which one he had in mind! In his view Quine's arguments rest on obviously false behavioristic premises and have had effects on the philosophy of language that are unwarranted and deleterious.

My own assessment of the situation differs quite a lot from this one. I am not convinced that there is a genuine conflict between Quine's critique of the analytic–synthetic distinction and any role it might have, or come to have, in the cognitive sciences. On the contrary I think we can find a formulation of Quine's position that is (a) close to what he says, (b) true, (c) philosophically significant, and (d) perfectly consistent with scientific results and prospects.

A nice question regarding the interpretation of Quine's position concerns the *degree* of his scepticism about meaning. Is he saying that "is analytic" and intimately related expressions such as "is synonymous with", "means that", "is the correct translation of", and so on are

(i) meaningless

(ii) never determinately applicable

or merely

(iii) determinately applicable much less often than we had thought?

In my opinion nothing stronger than the third and weakest of these theses can be found in Quine's writings or justified by his arguments. For there are various places at which determinate semantic facts are acknowledged:

(A) Quine allows, both in "Two Dogmas"[3] and elsewhere, that there exist *stipulative definitions* which generate sentences that —at least at the very moments of stipulation— cannot be denied without violating their meanings. For example, when it was first decided to give the name "deuterium" to the atom with one proton and one neutron, then,

[3] W.V. Quine, "Two Dogmas of Empiricism" in his *From a Logical Point of View* (Cambridge, Mass.: Harvard University Press, 1953).

at that time, the sentence "Deuterium is an atom with one proton and one neutron" was determinately analytic.

(B) He is prepared to define the *analytic* truths as those reducible, via substitution of synonyms, to *logical* truths. Thus "All bachelors are unmarried" will be designated analytic if it is thought to be reducible in this way to "All unmarried men are unmarried". In that case, whatever difficulty may emerge with the notion of *synonymy*, at least the *logical* truths will qualify as determinately analytic.

(C) The thesis of Chapter II of *Word and Object*[4] is that there is no determinate fact of the matter as to which, among many radically different translation manuals, is the correct one. This does not imply that just *any* old manual might be correct. There are, Quine says, *some* objective standards of correctness, and these will rule out certain manuals. In particular, a manual that inter–translates a pair of observation sentences that are affirmed in different stimulus conditions is objectively incorrect. "Gavagai" might mean "Rabbit" or "Undetatched bit of rabbit", but definitely not "table".

Thus it seems clear that Quine's thesis falls far short of a thoroughgoing scepticism or eliminitivism about determinate semantic facts. What he is claiming is that such facts are surprisingly few and far between.

But surely, it will be said, there must be more to Quine's famous and influential critique than this far from earth–shaking point —and I agree. The real substance and philosophical bite of his discussion of semantic notions lies, not so much in their characterization as riddled with indeterminacy, but rather in their removal from epistemology: that is, in the view that such notions as 'true in virtue of meaning alone' have been overly relied on in philosophy and play no evident role in theories of the ways in which our beliefs are acquired and justified.

Remember that at the time Quine began his critical scrutiny of the analytic–synthetic distinction, this distinction was a central notion in the *reductionist* epistemological theory of

[4] W.V. Quine, *Word and Object* (Cambridge, Mass.: MIT Press, 1960).

the logical empiricists and positivists. This approach, pioneered by Frege[5] for the case of arithmetic, extended to the empirical sciences by Russell,[6] and elaborated in detail by Carnap,[7] was in full swing. My present point is that it was to an analytic–synthetic distinction playing the epistemological role to which it was assigned within logical empiricism —and not to the analytic–synthetic distinction *per se*— that Quine was mainly objecting. The essence of his objection was that the reductionist form of logical empiricism is incorrect and that it should be abandoned in favor of a different epistemological model —the holistic web–of–belief— a model in which semantic notions such as the analytic–synthetic distinction do not appear.

Let me be a bit more specific. A leading idea of logical empiricism was to separate science from metaphysical mumbo–jumbo, and to place it on a secure epistemological foundation. This was to be done through the supposition that each scientific hypothesis has empirical content. And this supposition was to be sustained by showing how each one could be deduced, with the help of appropriate definitions, from a combination of indubitable facts of experience and self–evident principles of logic. But how could one be really sure of the correctness of the allegedly self–evident definitions and logical principles that are needed? The answer favored by Carnap, Schlick, Ayer, and the rest of the positivists in the 20's, 30's and 40's, was that these sentences are *stipulations*, helping to fix the meanings of their constituent expressions; so their correctness is guaranteed by what they mean; they are analytic.

Thus the reductionist epistemological model contained the following elements:

(a) A determinate body of stipulated analytic sentences

[5] G. Frege, *The Foundations of Arithmetic* (1884) (Evanston, Ill.: Northwestern University Press, 1980).

[6] B. Russell, "The Relation of Sense Data to Physics" (1914), in *Mysticism and Logic* (London: George Allen and Unwin, 1963).

[7] R. Carnap, *The Logical Structure of the World: Pseudoproblems in Philosophy* (1928), trans. by R.A.George (Berkeley and Los Angeles: University of California Press, 1967).

defining logical, mathematical and theoretical terms, and specifying the empirical contents of individual scientific hypotheses;

(b) Epistemological decisions concerning which such hypotheses to accept in light of observation;

(c) Pragmatic decisions concerning whether or not to adopt new stipulations (that is, whether to speak a new language).

And these elements were deployed, first, to mark off the class of meaningful hypotheses, and second, to explain the dynamics of scientific theory formulation.

Notice that it was already a feature of this model that no sentence is immune from revision (i.e. removal from our system of belief–expressions). Contrary to what is often supposed, this thesis was not one of Quine's innovations. What was new with Quine was to deny (what the reductionist asserts) that whenever a sentence is revised there is a determinate fact of the matter as to whether this change does or does not imply a change of language, and there is a determinate fact of the matter as to whether our decision to abandon the sentence is epistemological or pragmatic. Quine's insight, in a nutshell, was that no explanatory work is done by supposing that there are such determinate facts of the matter; so we can perfectly well drop that assumption. For, in the first place, the model of belief–change that remains (whereby we move conservatively towards the globally simplest theory–formulation that accommodates our data) is both adequate and uncontroversial. And in the second place, the criterion of scientific respectibility wanted by the empiricists —the mark of intelligibility as opposed to mumbo–jumbo— need be nothing more than credibility within this holistic model; the requirement of individual empirical contents was unnecessary.

Thus Quine's primary claim is that reductionism should be rejected. Not only does it call for various determinate distinctions that resist characterization and seem not to exist, but, as we have just seen, its explanatory work is done better by the holistic web–of–belief model in which those distinctions are not employed.

Supposing that this critique of reductionism is correct, then even if cognitive science does (or will) find it explanatorily worthwile to postulate fully determinate semantic properties, including determinate analyticity, this can have no philosophical import. Granted, if we could determine by some sophisticated scientific technique which sentences were true in virtue of their meanings alone, then we could determine which changes in theory–formulation required a change in our language, and we could determine which changes in theory–formulation were epistemological matters and which were merely pragmatic. But would any of this be a good thing? What would it be good for? Quine's view of how science evolves, which seems more–or–less right, is that we proceed by moving conservatively towards the globally simplest theory–formulation that accommodates our data. What matters in science is predictive and explanatory success, and we simply don't care about the extent to which we are changing the meanings of words and the extent to which our decisions are pragmatic. Therefore, as far as science and the philosophy of science are concerned, a refined ability to detect instances of analyticity would not be useful. The situation parallels other areas in which a high degree of indeterminacy exists and is innoccuous. Vague concepts such as 'near', 'big', 'red', and 'old', do their jobs perfectly well despite the numerous cases in which their application is indeterminate. And unless our needs change, there is no point in making them more precise. Similarly, no philosophical benefit can accrue from the scientific discovery of determinate semantic properties.

So far I have discussed two closely affiliated respects in which Quine's critique of the analytic–synthetic distinction would be left undisturbed by its discovery by linguists or psychologists. It would remain true, as he said, that the reductionism which called for the distinction is a bad theory. And it would remain true that a scientific analytic–synthetic distinction would have no philosophical significance. However, Quine did more than question the epistemological value of determinate semantic facts. He denied that there were any (or at least *many*). Isn't *this* component of his view jeopardized by current cognitive theory?

It seems to me that Quine has two potential lines of defence against this criticism. In the first place, it might be said that he never went so far as to *assert the non–existence* of a determinate property of analyticity, but maintained merely the weaker thesis that the widespread belief in such a thing was unjustified —a *dogma* of empiricism, an article of faith, as he put it. In that case Quine's position is consistent with the subsequent discovery of empirical reasons for acknowledging that property.

The trouble with this defence is that Quine *does* argue for, and commit himself to, the stronger thesis —perhaps not early on, in "Two Dogmas", but certainly in *Word and Object*. The message of that work is that most facts of translation are heavily indeterminate, and that most facts of intra–linguistic synonymy and analyticity are also heavily indeterminate. These conclusions allegedly follow from the alleged fact that language is a collection of behavioral dispositions. And these conclusions certainly seem to be in sharp conflict with the hypotheses of determinate semantic properties which have proved (or at least *could* prove) explanatorily valuable in cognitive science. One might well feel drawn to Chomsky's view that these scientific results provide a *reductio* of Quine's behavioristic premise.

An alternative assessment, however, which I think worth considering, is that the apparent conflict between Quine's claims about meaning and the claims of cognitive science about meaning may be dissolved by recognizing an ambiguity in the word "meaning".

It often happens that a word in everyday use is commandeered by scientists and construed in the context of their theories somewhat differently from the way it is normally understood. It is given a technical meaning, bearing some affinity to the meaning in ordinary language, but not exactly the same as it. This sort of thing is not a surprising or mysterious phenomenon; it has happened to the words "fish", "heat", "light", "line" and "solid"; and it could well happen to the word "meaning" itself. That is to say, the word "meaning" has a certain meaning in ordinary English (when, for example, one says "What is the meaning of the Spanish word 'cerveza'"); but it might well have a different meaning in the context of a scientific theory of language. Now it seems to me that this is not merely an abstract possibility. I think that the ordinary notion of meaning should certainly not be identified with the notion expressed by that word in, for example, Chomsky's 'internalist' theory of language.[8] If this is so, then one can begin to see the possibility of reconciliation between Quine, who holds that meaning is heavily indeterminate, and the advocates of a theory of language in which indeterminacy is denied. We can say that Quine was talking about the ordinary notion of meaning, and that he is not contradicted by claims regarding scientific–meaning, a quite different phenomenon.

But what grounds are there for supposing that the ordinary and the scientific uses of the word "meaning" refer to distinct phenomena? As it stands, this may look to be a bad question. For there is no such thing as *the* scientific use of the word "meaning": different theories within linguistics and psychology use the word in very different ways. And so it may seem to be impossible to show that there could be no scientific theory in which the word "meaning" refers to what we ordinarily refer to, and which allows for a determinate specification of what is "true in virtue of meaning". In fact, however, it is not hard to make something of a case that whenever an ordinary predicate is comandeered by science it suffers a change of meaning and comes to express a different property.

Consider, for example, the expression, "old man". This has a function in ordinary language. Suppose biologists discover an important physiological change that takes place in different men at different ages and that is connected with the familiar manifestations of aging. And suppose that within their theory of this phenomenon they come to use "old man" as a technical term for any man who has undergone this physiological change. Will this be evidence that the expression,

[8] N. Chomsky, *Knowledge of Language* (New York: Praeger, 1986) and *Language and Problems of Knowledge: The Managua Lectures* (Cambridge, Mass.: MIT Press, 1988).

"old man", is not really vague, as we had always thought? Will we have discovered that some teenager, who by a freak accident has undergone the change, is really an old man? Obviously not. Rather, the predicate, "old man", will express one property for the biologists and a different one for us. We say this because we can see that the scientists have given the original expression a new function and have equipped it to perform that function by assigning it some new rules of use.

Similarly, it may be said on behalf of Quine —and with considerable plausibility, I think— that the expressions, "is synonymous with", "is a truth of language", "is the correct translation of", and so on, have a function in ordinary life, which they perform perfectly well despite their indeterminacy; and that when a scientist comes along to press these expressions into new explanatory service, their uses, their meanings, and the properties they express, are changed. The new properties may well be determinate; but this will not contradict Quine's thesis that the semantic properties to which we ordinarily refer are indeterminate.

Suppose, for example, a theorist of language (say, Jerry Katz[9]) discovers that the meaning of each expression has a hidden structure: in other words, it becomes explanatorily valuable to postulate such structures. Suppose that it is then possible to define a relation, R, between expressions in terms of some theoretical relation holding between their underlying structures. And suppose that there is a high correlation between instances of this relation and 'intuitive analyticity': i.e. often, when "A" stands in relation R to "B", then "All A's are B" is felt to be analytic, and vice versa. Now suppose that, for this reason, our theoretician designates a sentence as "really analytic" just in case it has the form "All A's are B" and "A" stands in R to "B". In that case it seems to me that the property to which the scientist refers is not the same as the one philosophers have had in mind. The term's use within the theory would be significantly different from, and could not be explained in terms of, its prior philsophical use.

---

[9]See J.J. Katz, *The Metaphysics of Meaning* (Cambridge, Mass., MIT Press, 1991).

One might well object, on grounds of perversity, to the line of defence I am proposing. For in defending Quine's denunciation of the concept of meaning I rely on that very concept —maintaining that the word "meaning" suffers a determinate change in meaning when it is employed by scientists. To this objection I have two responses.

In the first place, as I said at the outset, I do not think that Quine's scepticism about semantic facts is as thoroughgoing as it is often taken to be. I read him as claiming that semantic properties are much less determinate that people (especially the logical empiricists) had realized —and not that there are no such properties. I certainly don't think that he feels prohibited from ever maintaining that two words have different meanings, or that expressions are sometimes ambiguous. And the view of the word "meaning" that I am attributing to him is simply a claim of this form. Nevertheless, I must concede that my defence of this claim is not solid: I cannot *prove* that any scientific, determinate concept of meaning would have to differ from the ordinary one.

However —and this is my second point— I should emphasize again that the denial of analyticity is not what is philosophically important in Quine's critique. His primary thesis, as I said above, is the epistemological irrelevance of analyticity; whether it exists or not is a secondary matter. As Quine himself now says:

> I now perceive that the philosophically important question about analyticity and the linguistic doctrine of logical truth is *not* how to explicate them; it is rather of their relevance to epistemology. The second dogma of empiricism, to the effect that each empirically meaningful sentence has an emprirical content of its own, was cited in "Two Dogmas" merely as encouraging false confidence in the notion of analyticity; but now I would say further that the second dogma creates a need for analyticity as a key notion of epistemology, and that the need lapses when we heed Duhem and set the second dogma aside.[10]

[10] W.V. Quine, "Reply to Hellman" in L. Hahn & P. Schillp (eds.), *The Philosophy of W.V. Quine* (La Salle, Ill.: Open Court, 1986).

In conclusion, let me summarize my main points. I have been arguing that the apparent conflict between Quine's critique of the analytic–synthetic distinction and its justified postulation within cognitive science is an illusion. For the following central features of Quine's position are unaffected by the scientific situation:

(1) that reductionist epistemology, which demanded a determinate analytic–synthetic distinction, is no good, and should be replaced by the holistic, web–of–belief model;

(2) that a scientific, fully determinate, analytic–synthetic distinction would have no philosophical import;[11]

(3) that the semantic properties to which we ordinarily refer may be indeterminate, even if those postulated within cognitive science are not.

It is a familiar (if often neglected) meta–philosophical dictum that philosophical considerations should not be allowed to constrain the direction of scientific research. The old a priori complaints about, for example, unconscious mental states, curved space, action at a distance, and backwards causation, stand as an embarassment to philosophy; for they rested on a simple failure to appreciate that technical notions were in play. And, of course, the same goes for analyticity: Quine's critique must not be allowed to inhibit the postulation of analogous properties within linguistics and psychology. However, the general moral of this paper is that we should recognize that, not surprisingly, the irrelevance of philosophy to science goes the other way too: scientific results won't solve philosophical problems.[12]

---

[11] I say "no philosophical import" instead of "no epistemological import" in order to leave room for the possibility that a scientific analytic–synthetic distinction would play some role in the causal explanation of belief, and would thereby figure in 'epistemology naturalized'.

[12] I would like to thank Ned Block, Paul Boghossian, Noam Chomsky, Martin Davies and Jerry Katz for helpful discussions of the material in this paper.

6

PHILOSOPHICAL ISSUES, 3
Science and Knowledge, 1993

Cognitive Science and the Analytic/Synthetic Distinction: Comments on Horwich[1]

Paul A. Boghossian

1. Quine is usually read as arguing either for a *non-factualism* about analyticity:

(1) The property of being true by virtue of meaning alone does not make coherent sense.

Or, at the very least, for an *error* thesis about it:

(2) Although the idea of an analytic sentence does make some sort of sense, there aren't and couldn't be, any instances of it.

---

[1]The present paper is a comment on Paul Horwich's "Chomsky and Quine on the Analytic–Synthetic Distinction", *Proceedings of the Aristotelian Society*, 1991. It is that paper that Horwich read at the meeting in Salamanca and on which I commented. Unfortunately, it proved impossible to publish that paper in these proceedings; what appears instead is a closely related companion piece. My comments remain largely unchanged.

These attributions —including the stronger non–factualist thesis— seem licensed by many passages, including the famous one which concludes Quine's discussion in "Two Dogmas":

> ...for all its a priori reasonableness, a boundary between analytic and synthetic statements simply has not been drawn. That there is such a distinction to be drawn at all is an unempirical dogma of empiricists, a metaphysical article of faith. (p. 37)

2. Nevertheless, Paul Horwich does not wish to read Quine as endorsing either (1) or (2). He certainly does not wish to attribute (1) to him. And he wishes to attribute only a *restricted* version of (2). What this restriction is and why Paul wants to make it are matters I'll come to in just a moment.

3. Before going on, however, I wish to point out that it strikes me as a *very good thing* for a philosopher to distance himself from claims (1) and (2), especially if he is not simultaneously prepared to subscribe to *meaning–indeterminacy* —i.e. to the view that there are no determinate facts about what linguistic expressions mean— and to its natural accompaniment, *meaning–eliminativism.*

Quine, famously, traveled the road from a denial of the analytic/synthetic distinction (in "Two Dogmas") to belief in meaning–indeterminacy and ultimately meaning–eliminativism (in *Word and Object*).

One of the most puzzling aspects of the contemporary reception of Quine, however, has been that although philosophers have gone along overwhelmingly with the denial of the analytic/synthetic distinction, they have not necessarily been willing to follow Quine to meaning indeterminacy and meaning–eliminativism. That is, they have thought of the denial of a/s as an *insight* into determinate meaning facts, rather than a rejection of them.

However, this strikes me as an impossible combination of views. Let me briefly indicate why.[2]

---

[2]The argument here is spelled out in detail in my "Analyticity", forthcoming in Hale and wright (eds.): *Blackwell's Complanion to the Philosophy of Language.*

Suppose meaning facts are determinate, so that there is an objective, determinate fact about what expressions mean. To make things vivid, imagine that the set of all possible concepts is arrayed neatly in Plato's heaven: $C_1, C_2, C_3, \ldots$ etc...And that for any expression, '*A*', '*B*' etc..., there is a fact of the matter about which of these various concepts it expresses. (This picture does not, of course, rule out *vague* concepts.)

Then, how could there fail to be a fact of the matter about whether any two expressions do or do not mean the same? By hypothesis, there is a fact about what concept '*A*' picks out and what concept '*B*' picks out.

Now, if they do pick out the same concept, then (assuming '*A*' and '*B*' are of the appropriate grammatical category) $A = B$ will express an analytic sentence.

Or, if you think for some or other reason that we can't use the identity sign as we actually have it for these purposes, we can imagine that there might have been a connective % whose meaning was precisely as follows: Sentences of the form $A \% B$ are true if and only if '*A*' and '*B*' have the same meaning, false otherwise. Given that we are operating on the assumption of determinate meaning facts, we are entitled to assume that there might have been such a connective. Similarly, we are entitled to assume that there might have been synonyms. The upshot is that, if meaning realism is true, I don't see how either of Quine's skeptical theses about analyticity could turn out to be true.

4. So, my problem with Quine's thesis (2) —I shall henceforth concentrate only on this weaker and hence more plausible thesis— is that it leads to the very radical doctrine of *global indeterminacy*. And I don't think either that global indeterminacy is true or that Quine's arguments for it are very compelling.

Paul has a different, if related, worry about (2) —the view that there couldn't be sentences that are true or false purely by virtue of their meaning. He doesn't want to attribute an unrestricted acceptance of (2) to Quine because he believes that (2) is in fact denied by contemporary empirical theories of linguistics and psychology. So Paul sets himself the task of answering the question:

> What thesis should we attribute to Quine which would both preserve the spirit of Quine's critique of a/s and which would not be falsified by cognitive science?

The answer he proposes is interesting and elegant. The main idea, as I understand it, is this: The sentences we assert have *two kinds of meaning*. On the one hand, they have the sort of meaning that we attribute to them ordinarily, in our normal folk–psychological idiom. According to Paul, this sort of meaning —he calls it *public language meaning*— is a *purely behavioral* property. Attributions of it presuppose nothing about the mind or brain which underlies linguistic activity. (p. 2)

On the other hand, they have the sort of meaning that is attributed to them in cognitive science —what Paul calls *I–language meaning*. And, according to Paul, this sort of meaning isn't a purely behavioral property: attributions of it may, typically will, take into account facts about the cognitive architecture and organization of the mind/brain which underlies linguistic activity.

With this distinction in place, Paul argues as follows:

(3) There can be no analytic sentences defined over public language meanings.

(4) There *can* be analytic sentences defined over I–language meanings.

(5) Only the sort of analyticity defined over public–language meaning can be put to the sort of epistemological use that Quine found objectionable. Hence, the fact that (2) is not true *simpliciter* doesn't affect any real interest of Quine's.

5. There's a lot to discuss here. In this space, I will only have time to concentrate on the first and most important claim —that the ordinary folk–psychological notion of meaning won't support the existence of sentences that are true or false purely by virtue of their meanings. What is the argument for this claim?

The basic idea seems to be this: Public–language meanings are purely behaviorally constituted. But the totality of the behavioral facts involving a person's use of a word don't settle what he means by that word, what rule of use is governing his use of the word. Hence, if that person's use of the word *changes* (think here of Paul's example involving the sentence 'Distinct straight lines don't intersect') then there won't necessarily be a fact of the matter as to whether this change in use reflects a change in the meaning of the word — a change in the rule governing it use— or whether it merely reflects a change in some *empirical belief*. Hence, there isn't a distinction between analytic and synthetic sentences.

In short: the argument is that behaviorally consituted meanings are *indeterminate*; hence, since public–language meanings are purely behaviorally constituted, they cannot support objective synonymies or analyticities (see p. 6).

Notice, first, that this reasoning doesn't flout the claim I was outlining above, that a denial of the a/s isn't consistent with meaning determinacy. On the contrary, the failure of an a/s distinction is traced precisely to an alleged failure of determinacy. However, Paul is not very expansive about the grounds for believing the major premise —that behaviorally constituted meanings are indeterminate. The argument here (pp. 5-6) goes by very quickly. Essentially two points are made.

The first is that since confirmation is holistic —since sentences don't have individual verification conditions in isolation from the overall theory in which they are embedded— a *verificationist* about meaning won't be able to characterize the notions of synonymy and analyticity.

One problem with this point is that even if it were it true would apply only to a verificationist theory of meaning, and that is not a particularly plausible view. The second problem is that it doesn't seem to me to be true in any case. A verificationist could say that *all* of the inferential liasions postulated by a theory for a sentence are consitutive of its meaning, so that *any* change in the theory would lead to changes in the meanings of all the constituent expressions. This may not be a very plausible analytical synthetic distinction, but plausibility wasn't the point at issue. If it were, we wouldn't be discussing verificationism in the first place.

Paul's second point is that Quine's web–of–belief model eschews semantical notions like analyticity, thus suggesting that they play no explanatory role in understanding the evolution of science.

I must say that I don't really see what's going on here. To the extent that the web–of–belief model is understood as already *incorporating* the idea that nothing is true by virtue of meaning alone, then it already *presupposes* the denial of the a/s distinction and so cannot be viewed as an argument for it.

So, I don't really see that we've been given a good argument for the claim that behaviorally constituted meanings are semantically indeterminate.

6. However, even granting this point, it seems to me that there is very little reason to believe the minor premise of Paul's argument —namely, that the meanings attributed in folk psychology are purely behaviorally constituted properties, properties whose attribution presupposes nothing about the character and nature of the mind/brain that underlies linguistic activity.

There are many things that could be said here, but let me confine myself to this. Given the singular lack of success of behaviorist accounts of our other folk–psychological concepts, it would be very surprising indeed if behaviorism were true about our folk–psychological concept of *meaning*. At any rate, the *argument* for the claim that it is such a notion would have to be very strong, if it is to overcome an understandable resistance.

Paul's argument for the claim that ordinary public language meaning is of a familiar type; it can be found in the writings of Dummett, and others in the Wittgenteinian tradition.

The idea is that meanings (and other folk psychological notions) must be notions of behavioral properties, else how are we to explain the fact that we so confidently infer what they are from the mere observation of a person's behavior.

> The possession of these [meaning properties]... is confidently inferred from everyday behavior. We observe the manner in which Mary uses words and the manner in which she responds

> to the use of words by others. On this basis, without any esoteric source of information about her mind or brain, we unhesitatingly conclude that she [speaks] English. (p. 2)

I have two difficulties with this sort of argument. In the first place, I think that it is possible to give a satisfactory account of the epistemology of other minds, one that would explain the confidence and ease with which we attribute folk–psychological properties, even if those properties are construed mentalistically.

Although this is obviously not the time to go into this in detail, and although it could benefit enormously from a more a refined treatment than it has hitherto received in the literature, the rough idea is familiar: We justifiably infer to the properties of another person's mind from the observations of their behavior and by analogy from our own case.

In the second place, what arguments like Paul's tend to underestimate, it seems to me, is the extent to which behavioral properties are not *themselves* easily observable properties. In particular, those of my behavioral properties that are the only remotely plausible candidates for constituting what I mean are not easily observable.

The point is that there is no way to reconstruct what I now mean by an expression from *my actual behavior with it right now.* That behavior, as Kripke's famous discussion of the concept of addition vividly shows, is consistent with my following an infinite number of rules.[3]

Hence, at a minimum, the behavioral properties that are to constitute meanings, on a behavioral view, must be highly dispositional, counterfactual and idealized.

Now we may ask: How much do you really know by *mere observation*, about a person's behavioral dispositions, with respect to expressions?

The answer is very little. Think again here of Kripke's famous discussion of '+'. Any bit of my behavior that you may have seen conforms to a large, indeed infinite, number of

---

[3]See Saul Kripke: *Wittgenstein on Rules and Private Language*, Harvard University Press, 1982.

dispositions. Which disposition I have is clearly not a merely observational matter.

And what this shows is that even a behaviorist, by his own lights, will have to use the sort of analogically mediated inference that I described the mentalist as exploiting, if he the behaviorist is to have an account of the epistemology of other minds.

But, then, the main motive behind being Behaviorist about public–language meaning disappears. There no longer remains any real reason for thinking that public–language meaning is behaviorist. And, hence, so far as any argument that either Quine or Paul has given us, there remains no reason to think that public–language meaning is indeterminate and, hence, no reason to think that it won't support objective synonymies or analyticities.

7. If what I've been saying is right, then the way is paved for a different and, I think, more satisfying picture of the relation between scientific investigation of meaning and the ordinary notion we attribute in common sense.

For Paul, it is crucial to claim that the notion of meaning studied in cognitive science is really a *different* notion from the one we ordinarily talk about, a change of subject matter. I think, on the contrary, however, that anyone not antecedently committed to a behaviorism about the ordinary notion would concede that it is heavily implicated in assumptions about the character and nature of the mental apparatus that underlies linguistic activity, assumptions that it is in part the proper business of cognitive science to investigate, refine and fill in.

In sum: I think Paul's attempt to reconcile Quine's rejection of a/s with the determinacy of meaning postulated by cognitive science fails. I think that Paul is right to think that Quine's rejection of a/s and his belief in indeterminacy rest largely on his behaviorism about the ordinary notion of meaning. But I think that behaviorism is wrong. I think the ordinary notion is mentalistic, determinate, supports an analytic/synthetic distinction and is much more smoothly related to the empirical study of meaning than Paul's picture allows.

PHILOSOPHICAL ISSUES, 3
Science and Knowledge, 1993

On the Logical Form of Propositions: Some Problems for Vanderveken's New Theory of Propositions

Jorge Rodríguez Marqueze

1 Introduction

The aim of this paper is to point out two problems affecting Daniel Vanderveken's new theory of propositions as it is presented in his recent book *Meaning and Speech Acts.*[1] These two difficulties are internal to the theory and have to do, on the one hand, with the kind of logical structure offered by Vanderveken as the logical type of propositions and, on the other, with what seems to be a basic epistemological notion of his theory: the notion of "apprehending" or "having in mind" or "fully understanding" a proposition.

---

[1] *Meaning and Speech Acts* (two volumes). Cambridge University Press, 1990. Numbers in parenthesis refer to pages of this book.

In this talk I shall proceed as follows: first, I shall give a general but selective summary of Vanderveken's theory (to this purpose, I shall make ample use of quotations or paraphrases of Vanderveken's own text) and, then, I shall indicate the two problems in question.

2 Vanderveken's Theory

Vanderveken's theory of propositions is part of a very ambitious project in semantics. His main goal in his aforementioned book is "to construct a general formal semantics for natural languages capable of characterizing the conditions of success as well as the truth conditions of literal utterances" (p. 2). The execution of this plan amounts, in the view of the author, to the attempt of unifying "speech act theory as developed in the tradition of Austin and Searle, and classical truth conditional formal semantics as developed in the tradition of Frege, Russell and Tarski" (p. 10).

The fundamental hypothesis underlying Vanderveken's entire project is "that complete illocutionary acts (and not only propositions or truth conditions) are the primary units of literal meaning" (p. 11). And since (elementary) illocutionary acts have the structure $F(P)$, where F is an illocutionary force and P is a proposition, Vanderveken's analysis of the logical form of propositions is one of the two basic elements (together with the analysis of illocutionary forces) of his analysis of illocutionary acts.

In accordance with the declared main purpose of his book, the author offers a theory of propositions which unifies a) the Fregean tradition, that understands propositions as senses of (utterances of) declarative sentences whose only specified condition of identity is identity of truth conditions, with b) the tradition of speech act theory that understands propositions as the contents of illocutionary acts.

Vanderveken's central assumption that propositions are contents of illocutionary acts imposes, among others, a crucial restriction on their logical analysis: "[I]dentical propositions must satisfy a criterion of substitutivity salva felicitate within illocutionary acts" (p. 81). That is, propositional identity cannot be reduced to identity of truth conditions because there are illocutionary acts with identical illocutionary force and strictly equivalent propositions as propositional content which still do not have the same conditions of success (they are not performed in the same contexts). For instance, an assertion that $2 + 3 = 5$ is not the same as an assertion that arithmetic is incomplete.

The author proposes to base the explanation of the failure of substitutivity salva felicitate of strictly equivalent propositions on "the fact that [propositions] are apprehended by speakers in their actual performances of illocutionary acts" (p. 82). And he further specifies that to apprehend a proposition is to understand "(1) the propositional constituents of that proposition, (2) the ways in which these propositional constituents relate in terms of predication in (...) the atomic propositions of that proposition, and (3) how the truth conditions of the complete proposition are determined from the truth possibilities of its constituent atomic propositions" (p. 85).

In Vanderveken's view, a definition of proposition which be in accordance with this characterization of the notion of apprehending a proposition must be one "that reflects that they are structured entities composed out of atomic propositions having propositional constituents of various types (such as relations, properties and senses of individuals) standing in certain logical relations of predication" (p. 86). A definition of this kind allows one to acknowledge that there are two, and not just one, essential components of each proposition: its truth conditions and also "the set of atomic propositions out of which it is composed" (*ibid.*) (i.e., the content of that proposition). And this brings as an immediate consequence that any adequate condition of propositional identity must take into account that two propositions may differ either because they have different truth conditions or because they have different contents. So, definitions of this type, unlike those whose only criterion of propositional identity is identity of truth conditions, avoid the familiar and bothersome consequence of having to identify all logical and mathematical truths as one and the same proposition.

Vanderveken's definition of propositions is, of course, a definition of this kind. He distinguishes between first order and second or higher order propositions (cf. infra). He offers first a definition of first order atomic propositions: "One can (...) identify in general semantics each first order atomic proposition with the ordered pair which contains first the finite set of its propositional constituents (these are an n-ary attribute of individuals and a certain number $k < n$ of individual concepts) and second the function which determines its truth conditions in accordance with the predication that is made with these propositional constituents" (p. 89). Next, he extends this definition for the case of second order atomic propositions. Finally, he gives a general definition for propositions of all kinds: "I will identify in general semantics each proposition with the ordered pair whose first term is the finite non-empty set of its atomic propositions and whose second term is the set of all truth value assignments to atomic propositions under which it is true" (p. 91).

According to Vanderveken's definition, strictly equivalent propositions which are composed out of different atomic propositions are not identical. This definition renders the intended result that strictly equivalent propositions may not be interchangeable salva felicitate: if two strictly equivalent propositions are composed out of different atomic propositions then, as in the case of the propositions that $2 + 3 = 5$ and that arithmetic is incomplete, it is possible that a speaker be only capable of "apprehending" the first but not the second. And this is enough to determine different conditions of success for, e.g., assertions with each of these propositions as contents.

3 First Problem

One of the problems affecting Vanderveken's theory is that the structure which he proposes for so called second order propositions (and in general, for higher order propositions) is not the one he needs in order to achieve what he declares to be the main goal of his analysis; namely, to reflect "what a speaker must have in mind when he expresses that proposition with an illocutionary force in an utterance" (p. 93). It is in order to serve this goal that Vanderveken formulates his complex criterion of propositional identity which requires not merely identity of truth conditions but also identity of content among propositions.

There is an unexplained (and —I think— unjustifiable) asymmetry between the treatment given respectively to first and second (or higher) order propositions. The latter are propositions, like (1) that being modest was a quality of Socrates and, more interestingly, like (2) that it is logically possible that John speaks French. (In general, propositions of second or higher order are, among others, all those expressed by utterances of sentences with non–truth–functional operators, like modal operators, and all those expressed by utterances of sentences with quantifiers; they are either propositions about propositions or propositions about attributes of immediately lower order.) At first sight, Vanderveken's characterization of second order atomic propositions seems to be relevantly similar to his characterization of first order atomic propositions: he identifies them with "ordered pairs whose first term is the set of propositional constituents (which contains in this case exactly one attribute of attributes of individuals) and whose second term is the function that determines the truth conditions" (p. 90). But, after a closer look at the constituents of second order atomic propositions, one discovers that there is a notable difference between them and the constituents of first order atomic propositions: First order atomic propositions refer or talk about individuals and have as constituents a) an attribute of individuals and b) individual concepts of the objects referred to. Second order atomic propositions are about first order propositions (or about attributes of individuals) and have as their constituents a) one attribute of propositions (or an attribute of attributes of individuals) and b) the very same propositions (or attributes of individuals) to which they refer. Indeed, Vanderveken analyses proposition (2) as a second order elementary proposition containing a single second order atomic proposition whose constituents are a) a modal property of propositions (x is true in some contexts) and b) the proposition that John speaks French. But the proposition that John speaks French is also the denotation or the reference of proposition (2), it is that which proposition (2) is directly about.

If the analysis of second order propositions were a straightforward projection of the analysis of first order propositions, there would be needed individual concepts of propositions, instead of the propositions themselves, and individual concepts of first order attributes, instead of the attributes themselves, as constituents of second order atomic propositions; and similarly for all higher order propositions.

This asymmetry is troublesome because almost exactly the same consideration which moves Vanderveken to choose individual concepts as constituents of first order propositions must be taken into account in the case of second order propositions: Namely, his interest in respecting the criterion of substitutivity salva felicitate as a criterion of propositional identity. Certainly, it may happen that two different second order propositions were about the same proposition but that they "presented" it under a different guise, in such a way that speakers may not be aware that the two propositions were about the same proposition. Think, for instance, of propositions (3) that it is possible that planet Pluto is inhabited and (4) that the proposition which is the content of the last utterance of yesterday's lecturer is possible, where the last utterance of yesterday's lecturer was that planet Pluto is inhabited. If identity of propositions means more than identity of truth conditions, as Vanderveken repeatedly emphasizes, then one needs individual concepts of propositions, and not the propositions themselves, as constituents of second order propositions. Vanderveken's analysis of second and higher order propositions commits him, in fact, to the acceptance of the theory of direct reference for all singular expressions referring to propositions and to the admission of singular propositions. But the author strongly emphasizes that his analysis of propositions is incompatible with direct reference and singular propositions, and he puts forward no less than four objections against them. As I show in another place,[2] of these four objections, only the first one is avoided by what

[2] Cf. "Proposiciones de orden superior como proposiciones singulares: una consecuencia indeseada de la teoría de la proposición de Vanderveken", in *Anales del VII Congreso de Lenguajes Naturales y Lenguajes Formales*, Vic, Barcelona, 1991 (forthcoming).

seems to be Vanderveken's view on higher order propositions. Therefore, there appears to be an inconsistency in the (literal) theory of propositions proposed by Vanderveken.

I do not know why Vanderveken proceeds the way he does with respect to higher order propositions. The only reason I can think of is that he wants to preserve some of his fundamental laws of illocutionary logic. For instance, the law of Elimination of Necessity: 'Necessarily *p*' strongly implies *p*'.[3] But, certainly, this is not a strong nor convincing enough reason.

4 Second Problem

Another problem affecting Vanderveken's analysis of propositions is that, when the analysis is applied to propositions expressed by sentences with indexical expressions, it turns out to be inconsistent with his principle of substitutivity for occurrences of sentences in contexts. This principle says that "sentences with non–synonymous clauses can express the same propositions in different contexts (if, for example, they differ only in that they contain non–synonymous indexical expressions at the same place, and these indexical expressions have the same semantic values in these contexts)" (p. 95, cf. also p. 48).

I shall argue that Vanderveken's analysis of propositions as ordered pairs of contents (sets of atomic propositions) and

---

[3]Indeed, suppose that, contrary to what is the case, Vanderveken's analysis of second and higher order propositions were a straightforward extension of his analysis of first order propositions; that is, suppose that the content of 'Necessarily *p*' were a) the predicate about propositions "*x* is true in all contexts" and b) the individual concept of *p* "being the proposition that *p*". Then, the content of *p* would not be included within the content of 'Necessarily *p*', since *p* has as its content a set of atomic propositions (composed each of *n* individual concepts and a *n*–adic predicate) neither of which are part of the content of 'Necessarily *p*'. But a proposition *A* strongly implies another *B*, according to Vanderveken's definition, iff i) the content of *B* is included within the content of *A* and ii) *A* strictly implies *B* (cf. p. 100). Therefore, 'Necessarily *p*' would not strongly imply *p*.

truth conditions does not leave any room for non–synonymous indexical expressions with the same semantic values. Or, more accurately, that only if one presupposes a notion of semantic value quite different from the one (which seems to be) adopted by Vanderveken it is possible to sustain the claim that there are (can be) non–synonymous indexical expressions with the same semantic value.

What does Vanderveken understand by semantic value of an expression? For the purposes of our analysis, and if we restrict our exploration to the case of those expressions which are either sentential clauses (indicators of propositional content) or categorematic components of sentential clauses,[4] it will suffice to mention three possible answers:

1) Semantic value of an expression in a context c = its extension in c.

2) Semantic value of an expression in c = its sense in c = a function from possible worlds into extensions.

3) Semantic value of an expression in c = its sense in c = an ordered pair whose 1st term is (roughly) the set of senses out of which the given sense is composed and whose 2nd term is a function from possible worlds[5] into extensions.

From Vanderveken's text (cf, e.g., pp. 42, 44 and 48), it is clear that his notion of semantic value is 3). But, while 1) and 2) are compatible with the view that there may be non–synonymous indexical expressions having the same semantic value, 3) is not. Let us see why this is so: That is, let us

---

[4] Hereafter, when using 'expression', unless explicit or contextual indication to the contrary, I shall be referring only to this restricted class of expressions.

[5] Vanderveken would talk of possible contexts of utterance, instead of possible worlds; but this only because he makes the common assumption "that each possible context of utterance takes place in one possible world" (cf. p. 80). For the purposes of my argument is totally immaterial whether we characterize the sense of an expression in terms of possible worlds, as, for convenience, I do it here, or in terms of possible contexts of utterance.

show, first, that Vanderveken's notion of semantic value is 3); and, second, that 3), contrary to 1) and 2), rules out the possibility of having non–synonymous indexical expressions with the same semantic value.

I shall take as paradigmatic of Vanderveken's notion of semantic value his notion of semantic value of a sentence. According to Vanderveken, the semantic value of a sentence S is the value of the function in which the linguistic meaning of S consists. The linguistic meaning of S is a function from possible contexts of utterance into illocutionary acts. Therefore, the semantic value of S in a context c is an illocuctionary act. This illocutionary act is the illocutionary act expressed by S in c. What is expressed by S in c is this illocutionary act. So, we can expect, in general, that what is expressed by a given expression in a context will be its semantic value. Now, an elementary illocuctionary act is logicaly composed of an illocutionary force F and a propositional content P. And the sentence S in c expresses an illocutionary act I because or inasmuch as it expresses the illocutionary force F of I and the propositional content P of I. One can distinguish two kinds of constituents (expressions) of a sentence from the point of view of the speech act theory: the indicator of illocutionary force and the indicator of propositional content; i.e., the sentential clause. The indicator of illocutionary force occurring in S expresses in each context the force of the illocutionary act expressed by S in that context. The indicator of propositional content of S (the clause of S) expresses in each context the proposition which is the content of the illocutionary act expressed by S in that context. Now, a proposition is an estructured entity and its structure is "reflected" somehow in the sentential clause which expresses it. The different constituents of the proposition P expressed by the clause of S in c are expressed in c by the different non–synonymous categorematic expressions which are part of the clause of S. So, the different components of P (attributes and individual concepts) are the semantic values in c of the categorematic expressions occurring in or as part of the clause of S. Therefore, according to this survey of Vanderveken's doctrine, the semantic value of a (categorematic) expression (occurring as part of an indicator of propositional content) is always a sense; i.e., it is always the value of a function which has (possible) contexts of utterance as arguments and senses (individual concepts, n-adic attributes of order n or propositions, depending on the kind of expression considered) as values.

This result allows us to disregard 1) as Vanderveken's notion of semantic value. The semantic value of an expression in a context is not its extension but its sense in that context. The question to ask now is: how does Vanderveken understand the notion of sense? We need to choose between the two alternative notions of sense presented, respectively, in 2) and 3). From Vanderveken's definition of propositions, it is perfectly evident that his notion of sense is the one described in 3). The notion of sense alluded to in 2) is no other than the standard notion of sense in possible–worlds semantics. But, as we saw at the beginning, it is an insatisfaction with that notion of sense (concretely, with that notion of the sense of sentences) which triggers Vanderveken's search for a new, more fine grained notion of proposition (sentential sense). 3) is simply a straightforward extension of Vanderveken's new notion of the sense of sentences (propositions) to the general case of the sense of all kind of (categorematic) expressions.

Why is the adoption of (the notion of sense presented in) 3) incompatible with the idea that there are (there may be) non–synonymous indexical expressions with the same semantic value? Because, if one applies 3) to the case of indexicals like 'I' and 'you', it turns out that their sense in a context is not merely a constant function from possible worlds (possible contexts) into individuals (cf. infra), rather, that constant function is only one of the terms of the ordered pairs in which the senses in a context of 'I' and 'you' respectively consist; the other term of the pairs being the set of senses out of which the senses of 'I' and 'you' in c are composed.

What are the senses out of which the senses (the individual concepts) of 'I' and 'you' in c are composed? I take for granted that, despite appearances, the senses (individual concepts) of 'I' and 'you' in a context are structured entities. That is, that the individual concepts expressed by 'I' and 'you' in c are, respectively, the individual concepts "canonically" expressed by 'the speaker in c' and 'the hearer in c'. But obviously, so understood, the individual concepts expressed in c by 'I' and 'you' are different structured entities, since they are composed of the senses of 'the speaker' and 'c', in the first case, and of the senses of 'the hearer' and 'c', in the second case (where, of course, the two occurrences of 'c' may refer to different contexts).

The sense of 'I' in a context c_1 is a certain individual concept, A, and the sense of 'you' in a context c_2 is also a certain individual concept, B. To claim, as Vanderveken does, that A and B may be the same individual concept provided that contexts c_1 and c_2 meet certain conditions; to claim this —I am arguing— amounts to understanding the notion of individual concept as the old, standard notion of sense of a singular expression in possible–worlds semantics: that is, as a function from possible worlds (contexts) into individuals. So understood, A and B would be the same in those cases in which the speaker of 'I' in c_1 where the same person as the hearer of 'you' in c_2. Indeed, the function from possible worlds into individuals expressed by a (possible) utterance of 'I' in c_1, where the speaker is John, is exactly the same as the function from possible worlds into individuals expressed by a (possible) utterance of 'you' in c_2, where the hearer is John. In both cases the function is constant: no matter what possible world (context) one takes as argument, the value of the function is one and the same individual, John.

Therefore, only if the notion of sense used by Vanderveken were the old, standard one of possible–worlds semantics, would it be the case that non–synonymous indexicals may have the same sense (express the same propositional content) in diferent contexts. And so, Vanderveken's claim about indexicals amounts, in fact, to denying or rejecting his own modification of the notion of sense for the specific case of (the individual concepts expressed by) indexicals. But the author cannot use the old standard notion of sense for singular expressions (for indexicals, in particular) while retaining or adopting his new, modified notion of sense for sentences. And this is so because there are cases in which the "difference in structure" which distinguishes two strictly equivalent propositions is, precisely, a difference in the structure (composition) of one or more of the ("strictly equivalent") individual concepts present, respectively, in each proposition. Thus, the propositions (5) that two plus three is a natural number and (6) that seven times nine minus fifty eight is a natural number are different but strictly equivalent propositions whose difference in structure is given or comes from (the different structure of) their different individual concepts: a) "being the number $2 + 3$" and b) "being the number $(7 \times 9) - 58$".

And similarly for the case of strictly equivalent propositions expressed by sentences containing non–synonymous indexicals: Consider the two following sentences: (1) I am Susana's friend and (2) You are Susana's friend. Suppose that (1) is uttered in a context c_1 where John is the speaker, and that (2) is uttered in a context c_2, in all similar to c_1 but for the fact that the speaker is Mary and the hearer is John. Bear also in mind i) Vanderveken's modified notion of sense as described in 3), together with these two tenets of Vanderveken's theory: ii) 'I' and 'You' are non–synonymous categorematic expressions. iii) "The categorematic expressions of a sentence are those which specify at each context a propositional constituent of a proposition which is the sense of that sentence in that context" (p. 85). By i), ii) and iii), 'I' and 'You' "specify" different propositional constituents of propositions in c_1 and c_2, despite the fact that they refer to the same object and despite the fact that the second term of the ordered pairs in which these propositional constituents (individual concepts) respectively consist is the same; and consequently, the propositions expressed by (1) in c_1 and by (2) in c_2 are also different. The expression 'I' has as its content in c_1, not just the person which happens to be the speaker in that context (John), nor the constant function from contexts of utterance to John, but the individual concept of that person which represents or identifies him as "the person who is the speaker in context c_1". It is this individual concept which is the content of "I" in the proposition expressed by (1) in c_1. Similarly, the content of 'You' in c_2 is not just the person who happens to be the hearer of (2) in c_2 (John), nor the constant function from contexts of utterance to John, but the individual concept of that person which identifies him as "the person who is the hearer of (2) in c_2". And the senses out of which the individual concept of being the speaker in c_1 is composed are obviously different from the senses out of which the individual concept of being the hearer in c_2 is composed. Different contexts of utterance may make (1) and (2) co–referential and, therefore, may make the propositions expressed by each strictly equivalent. But still, the two propositions will not and cannot be identical by Vanderveken's account, since the senses of 'I' and 'You' are always different and therefore the contents of the respective propositions have to be different too.

Notice, also and finally, that the (or a) crucial reason which moves Vanderveken to propose his modification of the notion of proposition —namely, his desire to be loyal to or to accommodate the (alleged) fact that propositions are not only the senses of sentences but also the contents of human thoughts and of illocutionary acts— is equally present or holds for the case of individual concepts expressed by indexicals in contexts. Thus, someone might have the belief that the hearer of (2) in c_2 is Susana's friend but not the belief that the speaker of (1) in c_1 is Susana's friend, etc., if, for instance, the person in question ignores the fact that the speaker of c_1 is the same person as the hearer of c_2.

8

PHILOSOPHICAL ISSUES, 3
Science and Knowledge, 1993

Mental Causation in a Physical World

Jaegwon Kim

1 Cartesianism and Mental Causation

The philosophy of mind as we practice it today is continuous with Descartes'. In particular, the metaphysical problem of devising an account of the mind–body relation which was the focal point of his concern has not really changed at all. The ontological picture presented to us by Descartes' dualism is familiar to us, at least in broad outlines: it divides the world into two independent domains of particulars, one consisting of mental substances and the other of material substances, and affiliates with each domain a distinct family of properties, mental properties for one and physical properties for the other. The substances and properties of each domain generate facts, events, states, processes, and the like, and these are given significant structure by the causal relations crisscrossing them. Each domain is conceived to be metaphysically independent of the other; the reality of one does not depend on that of the other —in current idiom, there are possible worlds in which only minds, but no matter, exist, and worlds in which only matter, but no minds, exist.

But Descartes wanted more. Unlike some of his contemporaries,[1] Descartes wanted the two domains in causal commerce with each other: although the domains are separate and ontologically independent, events in one can, and sometimes do, causally influence events in the other. In allowing causal linkages between the two domains, Descartes may appear to be only endorsing what seems so manifestly evident to common sense; moreover, as has often been remarked, our conception of ourselves as agents and knowers seems intimately tied to the possibility of causal influence between mind and body. For how would perception be possible unless external physical conditions causally influenced our sensory experience? How could there be human action unless our beliefs and desires, our decisions and volitions, had the power to move our limbs and thereby alter the arrangement of objects around us?

This apparently slight departure, seemingly so well justified, from the purer dualisms of Leibniz and Malebranche was to introduce an irresolvable tension into Descartes' scheme and prove to be its ultimate undoing. Allowing causal interaction between the two domains turned out to be a far more serious breach of the autonomy and independence of the two domains than might have been foreseen by Descartes and other interactionist dualists; opening this little channel through the dyke between matter and mind may well have in the end destroyed the dyke. The question how minds, as mental substances, could possibly influence the motion of physical particles was immediately raised by some of Descartes' contemporaries. In 1643, Princess Elisabeth of Bohemia wrote to Descartes, challenging him to explain "how man's soul, being only a thinking substance, can determine animal spirits so as to cause voluntary actions".[2] Ever since there has been a near consensus that the problem of

---

[1] E.g., Leibniz and Malebranche.

[2] *The Essential Descartes*, ed. Margaret Wilson (New York: New American Library, 1969), p. 373.

mental causation cannot be solved within the terms set by Descartes' substantival dualism.

Descartes' dualist interactionism has nevertheless enjoyed an enduring popularity; its intuitive appeal to common sense and its ability to fill some of our pressing conceptual and philosophical requirements are undeniable. Its appeal, however, has consisted not in its substance dualism —in particular, its reification of minds as substances— but in its twin claims that our mentality is distinguishable from the mere material aspect of our being, but that our mental life is causally intertwined with our material nature. The idea of mind as a substance has not been a serious contender in philosophy of mind for some time;[3] it was a good riddance, too, since its demise has undoubtedly removed one of the main obstacles to a satisfactory explanation of how psychophysical causation is possible. For, as Princess Elisabeth and many others wondered out loud, how could mental substances, without any spatial characteristics, causally influence bodies in physical space whose principle of motion was supposed to be impact action? How could thinking, an immaterial process in some immaterial substance entirely outside physical space, influence the motion of even a single material particle?

These questions are, however, not much more than expressions of vague worries and puzzlements, which may only elicit from the Cartesian the short conversation-ending retort "Why not?". I believe we can go further: we can show, I think, that, under Cartesian substance dualism, mental causation —that is, causal relations involving mental events— encounters an apparently irresolvable metaphysical difficulty ("the pairing problem", as I have called it elsewhere[4]). This

[3] For some current dissenting views, see Richard Swinburne, *The Evolution of the Soul* (Oxford: Clarendon Press, 1986), and W.D. Hart, *The Engines of the Soul* (Cambridge: Cambridge University Press, 1988).

[4] I discuss the pairing problem in a general setting in "Causation, Nomic Subsumption, and the Concept of Event", *Journal of Philosophy* 70 (1973): 217-236. Ernest Sosa has raised this issue specifically in connection with mental causation, in his "Mind-Body Interaction and Supervenient Causation", *Midwest Studies in Philosophy* 9 (1984): 271-282.

problem arises as follows: suppose that my mind's desire to raise my arm causes my arm to go up, and, simultaneously, your mind's desire to raise your arm causes your arm to go up. We have here two causes (i.e., two desires to raise an arm) and two effects (i.e., two arm risings): How do we know that the right cause has been paired with the right effect? What makes it the case that the desire in this mental substance was the cause of a change in this material substance, not of the same change in that other material substance? When we are dealing with changes in two bodies, the pairing is usually done by spatial relations: this flame causes this lump of wax to melt, and not that one, because this lump, but not that one, stands in an appropriate spatial relation to the flame (say, being within an inch from the top of the flame), and usually (as in this case) we can trace out a continuous causal chain between a cause and its effect. But no mental substance is in any spatial relation to a material body; remember that minds as substances have no spatial characteristics and no location in physical space. Moreover, the nonspatiality of mental substances makes it implausible to require a causal chain for every causal relation involving mental events (assuming that the idea of a causal chain makes sense for nonspatial events). So spatial relations cannot be counted on to generate correct pairings where mental substances are involved, and it is difficult to see what possible relations could give us a solution to this problem. Might one say that the "ownership" relation could supply the required pairing —that is, the desire in a given mind causes the change in that body which it "owns" (or with which it is "united")? But this would leave the relation of ownership mysterious and inexplicable, since the only intelligible account of it invokes the causal relation; my mind "owns" my body because it is the body with which it directly interacts.[5]

And note this important point: an essentially similar problem can be seen to arise for mental–to–mental causation, cases in which one mind is thought to have a causal influence on another mind. Suppose that a certain change in mind

[5] For further discussion of this issue, see Sosa, "Mind–Body Interaction and Supervenient Causation", pp. 274-275.

M_1 causes a change in mind M_2, and that, simultaneously, a change of the same kind in mind M_3 produces a similar effect in mind M_4. In virtue of what does the change in M_1 constitute a cause of the change in M_2 but not of the change in M_4? This question, I believe, is as difficult to answer as the question in the case of mental–to–physical or physical–to–mental causation.[6] It would seem then that substance dualism is going to have problems not only with causation between mind and body but with causation between one mind and another. If this is right, Descartes' doctrine of mind–body interaction should not be made to shoulder all the blame for the demise of mental substance. For the difficulty goes deeper; it appears that substance dualism does not allow a coherent notion of causation for mental substances.

Ridding ourselves of mental substance means that we now must make sense of mental causation in a fundamentally physical world. Since there are no mental particulars, mentality must now be understood as events, states, and processes in physical systems —complex physical structures, such as higher biological organisms— and not as happenings in some immaterial minds or spirits. In trying to understand how it is possible for our beliefs and desires to cause our limbs to move, we must take beliefs, desires, and the rest to be states and processes of our bodies. Perhaps, it ultimately makes no sense to think of nonphysical states as occurring to a physical structure; however, we are not assuming, at this stage, that mental events and states are *nonphysical* —we are only assuming that there are such events and states as beliefs, desires, emotions, sensations, and perceptions. In calling them "mental", we are not prejudging their nature, in particular their relationship to physical events and processes. They may yet turn out to be a subset of complex physical events and states.

---

[6] Perhaps, the mental–mental case is more difficult than the mental–physical case since for the former no relationship to material bodies is going to help with the pairing problem. The present difficulty appears to be related to the point that has been made by some philosophers to the effect that on the nonspatial conception of the mental the individuation of token mental states becomes problematic. See, e.g., Anthony Quinton, *The Nature of Things* (London: Routledge Kegan Paul, 1973), pp. 334-337.

2 The World as a Layered Structure

As we saw, Cartesian dualism presents the world as a bifurcated system constituted by two mutually exclusive and metaphysically independent, though causally connected, domains of entities. With the banishing of mental substance, the elegant bifurcated world was lost to us, leaving the physical world as the sole domain. It seems that we need to find structure somewhere, somehow, and it is no wonder that the Cartesian model was replaced by another one. This is what might be called a *layered* world, a hierarchically stratified system of "levels", or "orders", of entities and their characteristic properties.

Although this isn't required by the layered model as such, it is generally thought, perhaps on an empirical scientific ground, that there is a bottom level, one consisting of whatever microphysics is going to tell us are the most basic physical particles (electrons, neutrons, quarks, or whatever) out of which all matter is composed. And these particles, whatever they are, are characterized by certain fundamental physical properties and relations (mass, spin, charm, or whatever). As we ascend to higher levels, we find structures that are made up exclusively of entities belonging to the lower levels, and, moreover, the entities at any given level are characterized by a set of properties distinctive to that level (lower–level properties may apply to them as well). Thus, at a certain level, we will find lumps of H_2O molecules, with such properties as transparency, power to dissolve sugar and salt, its characteristic density and viscosity, etc. At still higher levels we will find cells and multi–cellular organisms with their "vital" properties, and farther up organisms with consciousness and intentionality. Beyond them, there are social groups of organisms, and perhaps groups consisting of such groups.[7] Sometimes, one speaks in terms of "levels of description", "levels of analysis", or "levels of language"; the layered model is often implicitly assumed in such talk.

---

[7] For a useful presentation of this layered picture, see Paul Oppenheim and Hilary Putnam, "Unity of Science as a Working Hypothesis", Herbert Feigl et al., eds., *Minnesota Studies in the Philosophy of Science*, vol 2 (Minneapolis: University of Minnesota Press, 1958).

Thus, the world as portrayed in the new picture consists of a hierarchy of levels, each level consisting of two components: a set of *entities* constituting the domain of particulars for that level and a set of *properties* defined over this domain. What gives this hierarchy structure is the mereological relation of *being part of*: entities belonging to a given layer are mereologically composed of entities belonging to the lower levels, and this relation generates an hierarchical ordering of the levels. As earlier noted, this multi–tiered picture usually carries with it the assumption that there is a bottom tier consisting of entities that are basic in that they lack physically significant proper parts.

This transition from the bifurcated world to the layered world fits in well with —perhaps, it has been necessitated by— the abandonment of mental substances. If mental substances are no longer available to serve as subjects of mental states and events, that is, as instantiators of mental properties, only physical structures are left to fill that role. Just as we attribute chemical and biological properties to complex structures composed of simple physical particles, we attribute mental properties to still more complex physical structures. Biological organisms are just complex physical structures that are wholly decomposable into simpler physical structures and capable of exhibiting biological characteristics; similarly, psychological creatures, or systems exhibiting mentality and consciousness, are biological (and perhaps nonbiological) structures with an appropriate level of organizational complexity. Thus, the renouncement of minds as immaterial substances seems to lead naturally to a layered view of reality.

One important aspect of the layered model remains to be specified: the relationship between the *properties* characteristic of things at a given level and the *properties* that characterize things belonging to its neighboring levels —that is, the inter–level relationship of properties. Given that things at distinct levels are ordered by the part–whole relation, should we expect that properties associated with different levels are also ordered by some distinctive and significant relationship?

It is not an exaggeration to say that this question has been at the center of the debate on the mind–body problem during the past few decades. The major rival positions on the mind–body problem that are being discussed today can be seen as alternative possible responses to that question. The classic positivist response was that the distinctive properties of a given level are *reducible to* the properties and relations of lower levels. That is *reductionism*. At first, meaning–preserving definition was taken as the primary vehicle of reduction, and that was *analytic behaviorism*; later, reduction was to be accomplished via empirically certified "bridge laws", and that was *type physicalism* which was influential in the 1950's and '60's. *Eliminativism* and *nonreductivism*, though united in their opposition to reductionism, differ on the significance of the failure of reduction. Eliminativism regards it as a sign of illegitimacy, holding that irreducible higher–level properties are "danglers" with no place in a true description of the world. Nonreductivism, on the other hand, considers irreducibility as a sign of autonomy: irreducible properties can form a perfectly legitimate autonomous set of scientific properties with genuine causal and explanatory powers, and must be included in any complete ontology of this world. *Emergentism* is a form of nonreductivism: it holds that such higher–level properties as mentality and vitality, though their emergence requires the presence of appropriate physical conditions, are irreducible to their "basal" physical conditions, and that these emergent properties possess the power for "downward causation", that is, the power to causally influence events and processes on the physical level. The doctrine of emergence was the first systematic articulation of the nonreductivist approach.[8] Most versions of functionalism, arguably the received position on the mind–body problem today, are nonreductivist, claiming that mental kinds as functionally understood are, on account of their "multiple realizability", irreducible to the underlying physical–biological kinds. (We will recur below to this antireductionist argument in greater detail.)

---

[8] For some recent discussions of emergentism see the essays in *Emergence or Reduction?*, ed. Ansgar Beckermann, Hans Flohr, and Jaegwon Kim (Berlin: Walter de Gruyter, 1992).

3 Microdeterminism: Supervenience and Physical Realization

The layered model, in which the world is portrayed as a system of layers of increasingly more complex material structures, is a picture that is grounded in, and promoted by, the modern physical and biological sciences. We usually associate modern physical science with the idea of *universal determinism*, the idea that no change or event takes place in this world except as a part of some lawful regularity involving antecedent events. This determinism is a principle of *diachronic determination*, from the past to the future; it is what underwrites the possibility of explaining the occurrence of events in terms of their causal antecedents. The layered picture of reality is often associated with another form of determinism which is orthogonal to diachronic determinism. This is a principle of *synchronic determination*: once all the facts about the basic particles of the world are fixed at any instant of time, that fixes all the facts about the world at that instant. That is, the world is the way it is at a time because all the basic particles are in a certain way at that time.[9] We can call this "microdeterminism". This is a special form of the supervenience thesis; it says that all the facts supervene on microphysical facts.[10]

I have put the thesis of microdeterminism "globally", in terms of worlds —"all the facts about the world", etc. But it can also be expressed "locally", thus: for any object, large or small, all of its properties are determined by its microphysical properties —that is, the properties and relations characterizing the basic particles it is composed of. This local version entails the global version; whether the converse entailment also holds can be debated.[11] However, we will

---

[9] Perhaps it is possible to formulate this thesis in a stronger form, to the effect that all the facts at any given level are fixed if all the facts at any of its lower level are fixed. That, however, would require a more detailed and precise description of the layered model.

[10] See Terence Horgan, "Supervenience and Microphysics", *Pacific Philosophical Quarterly* 63 (1982): 29-43.

[11] See Cranston Paull and Ted Sider, "In Defense of Global Super-

bypass this issue by adopting local microdeterminism as the primary thesis.

Microdeterminism does not exclude facts other than physical facts; rather, it holds that if there are facts other than facts about basic particles, for example, irreducibly biological or psychological facts, then those facts are determined by, and dependent on, microphysical facts. It is not for nothing, one might say, that we think of physics as the basic science. Microdeterminism says that microphysical facts are "the basic facts" on which all other facts depend.

Reductionism is committed to microdeterminism: if higher properties are reducible to lower ones, then what higher properties are instantiated in the world will simply be a question of what lower properties are instantiated. On reductionism, there really are no distinct "higher–level" properties or facts. But reductionism is not a popular option for most philosophers of mind today; it has not been one for at least two decades. However, contemporary nonreductivists are, by and large, physicalists and believe in the supervenience of all the facts on the physical facts.[12] Microdeterminism goes beyond the supervenience thesis, but the excess content is just the claim that all physical facts supervene on microphysical facts. This isn't something many people will dispute; in any case, it has no direct relevance to the mind–body problem.

The belief in the supervenience thesis and microdeterminism is widely held by the emergentists; in spite of their claim that higher emergent properties are not reductively explainable in terms of lower–level properties, they hold that emergent phenomena arise necessarily when, and only when, appropriate conditions are present at the "basal" level. As Lloyd Morgan, a leading emergentist said, emergentism explicitly rejected "any insertion into physicochemical evolution of an alien influence", such as entelechies, spirits, or

---

venience", forthcoming in *Philosophy and Phenomenological Research*, and further references therein.

[12] E.g., see Geoffrey Hellman and Frank Thompson, "Physicalism: Ontology, Determination, and Reduction", *Journal of Philosophy* 72 (1975): 551-564; John Post, *The Faces of Existence* (Ithaca: Cornell University Press, 1987). These authors, however, prefer the global version of supervenience.

elan vital.[13] Many philosophers talk of "physical realization" of mentality, rather than supervenience or emergence; in fact, the nonreductive form of functionalism is inseparable from the idea that very diverse physical structures can "realize" or "implement" a single mental property. Whether it is put in terms of realization, supervenience, or emergence, the core of the idea is the same: mental properties are instantiated in physical systems when, and only when, appropriate physical states occur in these systems. This implies the determination of the mental states of a system by its physical states.[14]

The problem of mental causation has thus been transformed into one of finding a causal role for mental events in this microdeterministic world, a layered world in which all that happens at higher levels (in particular, the psychological level) is wholly determined by what happens at the microphysical level. So the question reduces to this: Is mental causation possible in a microdeterministic world? Is there room in such a world for mentality to exercise its causal powers?

4 Why Nonreductivism Cannot Explain Mental Causation

What I will argue is that under nonreductivism the possibility of mental causation cannot be explained. This leaves reductionism and eliminativism as the remaining alternatives; I will be recommending, somewhat tentatively, a sort of hybrid solution that falls somewhere between reduction and elimination. But first we must attend to nonreductivism: my argument will make use of certain additional premises, which

[13] Lloyd Morgan, *Emergent Evolution* (London: Williams and Norgate, 1923), p. 15.

[14] Do considerations on "wide content" falsify a determination thesis of this kind? Not necessarily; what they show (although this is itself controversial) is that wide-content states are relational states, states involving relations to external objects and conditions. To accommodate them we only need to construe the underlying physical states relationally also. Details, however, are beyond the scope of this paper.

will be introduced at appropriate stages; these premises are inspired by physicalism and will not be rejected by those who accept the layered model.[15]

Let *m* be a mental event consisting in the instantiation of some mental property, M, at a time. To speak of the causal role or efficacy of *m* is in effect to speak of the causal role or efficacy of property M. If an instantiation of M is to have causal efficacy, it must do so by causing some other property Q to be instantiated. We can distinguish two cases: (i) Q is itself a mental property, and (ii) Q is a physical property. (i), therefore, is an instance of mental–mental causation; (ii) is mental–physical causation. I will first argue that (i) presupposes (ii) —that is, mental–mental causation is possible only if mental–physical causation is possible.

Suppose then that a given instantiation of M is causally efficacious in bringing about an instantiation of some mental property M^*. The nonreductivist who accepts the layered model believes in the physical supervenience, or realization, of the mental; for simplicity, let us adopt a single terminology, that of "realization" (details of the argument to follow depend, to some extent, on whether it is couched in terms of supervenience or realization; however, the main line of consideration remains the same in either case). The thesis that all mental properties are physically realized, therefore, is an additional premise of the argument. We may assume then that M^* is physically realized on this occasion by a certain physical property P^*, and this means that M^* occurs on this occasion in virtue of the fact that P^* is instantiated on that occasion. Thus, we have two apparently independent stories concerning why M^* is instantiated on this occasion: (1) Because an instantiation of M caused it; and (2) Because P^* was instantiated and P^* realizes M^*. There clearly is a tension here: if P^* indeed is a realization base of M^*, then no

[15] Arguments of this section are presented in greater detail in the following two papers of mine: "'Downward Causation' in Emergentism and Nonreductive Physicalism", in *Emergence or Reduction?*, ed. Beckermann, Flohr, and Kim; "The Nonreductivist's Troubles with Mental Causation", in *Mental Causation*, ed. John Heil and Alfred Mele (New York: Oxford University Press, 1992).

matter what else happens before the occurrence of P^* on this occasion, M^* would have occurred as long as P^* occurred. This makes unclear what role M can have in an account of the occurrence of M^*. On the other hand, if M truly is a sufficient cause of M^*, M^* must occur on this occasion no matter what else occurs; but ex hypothesi M^* cannot occur unless P^* occurs on this occasion. (We may assume, without real loss of generality, that there was no alternative realization base of M^* *on this occasion.*)

It seems clear that the only way we could reconcile the two stories, (1) and (2), is to postulate that M causes P^*, the realization base of M^*, thereby causing M^* to be instantiated on this occasion. This of course is an instance of mental–physical causation. I think this is a perfectly natural picture, and conforms to the following general principle, which I believe will be accepted by most contemporary physicalists:

> If a given instance of Q occurs in virtue of being realized by P (or in virtue of its supervenience on P), then an event is a cause of this instance of Q only if it is also a cause of this instance of P; and the only way of causing a realized (supervenient) property to occur is by causing one of its realization (supervenience) bases to occur.

This is especially plausible when you consider moral or aesthetic properties: to create beauty you must create a material object with a certain set of physical properties (think of creating a piece of sculpture) —there is no other way. And when the physical labor is finished, there is no further work you need to do; there is no further step of "creating beauty" that needs to, or can, be undertaken.

So the possibility of mental causation depends on the possibility of mental–physical causation. But can we make sense of this kind of mental causation under nonreductivism? I'll show why this is highly problematic.

Returning to our schematic example, then, we have it that an instance of M causes physical property P^* to be instantiated. But M is a psychological property and, therefore, must itself be realized by some physical property, P. We must now focus on the following claim: M is realized by P

and M causes P^*. (To be precise, "M", "P", and "P^*", as elsewhere, must be prefixed appropriately by "this instance of" and the like; for brevity, we use the abbreviated form.) The following question now arises: Why not consider P to be the direct cause of P^*, bypassing M altogether as an intervening link? Why not treat M as a mere epiphenomenon? What is gained by the insertion of M as a cause in this situation?

If we think of causation in terms of nomological regularity, there would be the required lawful regularity from P to P^*; the nomic requirement can be met. If we think of causation in terms of counterfactuals, we can fix the situation in such a way as to make the following hold: if P hadn't occurred on this occasion, M would not have, and hence P^* would not have either. It is in any case difficult to reject, on the layered model, the following principle:

> If property M is instantiated on a given occasion in virtue of being realized by (or in virtue of its supervenience on) property N, then the causal powers of M are identical with (at least, cannot exceed) the causal powers of N.

It says that there can be no excess causal powers magically accruing to higher properties that are not also among the powers of lower properties on which the higher properties depend.

What would happen if you denied this principle? Suppose M has a causal power, not had by any of its lower–level realization bases. As we saw, M's causal efficacy entails M's power to affect a lower–level physical property P^*. Since P^* is assumed to supervene on microphysical properties, changes in respect of P^* entail changes in respect of microphysical properties. This means, on the layered model, that when complex material objects with higher property M comes into being, certain microphysical properties begin to behave in ways different from the way they did before the emergence of M. Thus, we must conclude that the emergence of M, with its excess causal powers, leads to the violation of laws at the lower level! On this supposition, basic laws of physics can change when mentality emerges! This is the notorious "downward causation".

Such a view has been held by certain people, notably by the emergentists. But I do not believe it will be found acceptable by philosophers with a fundamental allegiance to physicalism and the autonomy of physics, and this includes almost all the philosophers currently engaged in the debate on the metaphysics of the mind–body problem. We must accept, then, the conclusion that whatever causal power this instance of M has is also had by the instance of its realizer P. And this seem to make M causally otiose: given that we have the P–instance serving as a cause of the P^*–instance, what additional causal work is left for the M–instance? Could we say that the M–instance is identical with the P–instance? But how could that be unless mental property M is identical with physical property P?

But such identities are precisely what the nonreductivist wishes to deny. It is clear that the difficulties will disappear if we accept standard reductionism —that is, type physicalism like the classic psychoneural identity theory— according to which every mental property is to be reductively identified with some underlying physical property. If mental property M is identical with physical property P, then in reality there is one property here, not two, and the causal powers of M are exactly those of P. And there is no special problem about accounting for the possibility of M's causing some property Q, beyond that of accounting for the possibility of P's causing Q. Under reductionism, mental causation turns out to be a special case of physical causation.

5 "Multiple Realization" and Reduction

But type physicalism has been abandoned by almost everyone, and is no longer a serious option for most philosophers. As is well known, the argument that has turned the tide against psychophysical reductionism was "the multiple realization argument" based on the observation that mental properties (in fact, all higher–order properties of the special sciences) can have diverse realizations in an extremely wide range of physical systems. The argument then goes on to assert that the nomologically possible physical bases of a mental property are so diverse and heterogeneous that they cannot be thought to constitute a unitary physical property that could serve as the reduction base for the mental property.[16] This shows, concludes the antireductionist, that mental properties are not physically reducible, and, further, that they must be taken as forming an autonomous domain for the irreducible special science of psychology.

Suppose mental property M is realized in physical properties P_1, P_2, and P_3 (the assumption of finitude is immaterial). Why not consider M to be reducible to the disjunction, $P_1 \vee P_2 \vee P_3$? The standard answer is that this disjunction is "wildly heterogeneous" and doesn't exhibit the kind of systematic unity we expect in a well-behaved property. I have argued elsewhere that there is reason to be cautious indeed in dealing with disjunctive properties; more specifically, I have argued that such disjunctions are likely to fail to be *nomic properties* capable of figuring in projectible laws.[17] I will not repeat that argument here, but the following line of consideration can be seen to point in the direction of such a conclusion. The idea that M's realizers are *diverse* is the idea that P_1, P_2, and P_3 are *diverse as causal powers*, and this means that the laws involving them are going to be very different (although in that they each realize M the causal structures involving them must show some sort of formal isomorphism). But treating their disjunction, $P_1 \vee P_2 \vee P_3$, as a single nomic property (representing a "natural kind") is to countenance the possibility of laws of the form $(P_1 \vee P_2 \vee P_3) \rightarrow Q$, where Q is some significant property. But a disjunctive law like this imputes the same causal powers to each of P_1, P_2, and P_3 in relation to property Q, since the law entails each of

---

[16] The argument originates in Hilary Putnam's "The Nature of Mental States", originally published under the title "Psychological Predicates" in W.H. Capitan and D.D. Merrill, eds., *Art, Mind, and Religion* (Pittsburgh: University of Pittsburgh Press, 1967). The present section makes use of material from my "Multiple Realization and the Metaphysics of Reduction", *Philosophy and Phenomenological Research* 52 (1992): 1-26.

[17] In "Multiple Realization and the Metaphysics of Reduction".

the following three laws: $P_1 \rightarrow Q$, $P_2 \rightarrow Q$, and $P_3 \rightarrow Q$. It plausibly follows then that if you take the disjunctive property seriously as a nomic property in your science, you are in effect treating the disjuncts involved as by and large sharing their causal powers. And this flies in the face of the assumption that M is multiply realized in heterogeneous structures, because diversity and heterogeneity here could only mean *causal* (or *nomic*) diversity and heterogeneity.

In any event, the standard antireductionist argument then proceeds to conclude that mental property M is not reducible to properties at the level of the P's. For none of the P's individually suffices to reduce M, and their disjunction doesn't qualify as a scientifically legitimate property; and no other possibility is in the offing. The antireductionist conclusion may seem assured.

But that isn't the only conclusion to draw. There is a more plausible alternative conclusion, and it is this: if the disjunction of the realizers of M cannot be regarded as a legitimate property on account of its failure to be nomic, nor can M be considered a legitimate, nomic property. That is, we need not, and perhaps must not, simply continue to protect our initial assumption that M is a well–behaved property, appropriately nomic and fit for scientific theory. That assumption is by no means sacrosanct; questioning the credentials of M as a property is one of the options open to us, and it is indeed the one I believe we should consider seriously.

The main reason for skepticism about the nomicity of M is the fact that M inherits causal inhomogeneity and unprojectibility from the heterogeneity of its realizers. For a quick argument for this conclusion, recall the principle in the preceding section about the causal powers of an instance of a multiply realized property M: an instance of M has exactly the causal powers of the realizing property involved in that instance. This means that the causal powers of M, or those of M–instances, are going to be as inhomogeneous as the causal powers of its realizers, P_1, P_2, and P_3, which, ex hypothesi, are extremely diverse and heterogeneous. More specifically, instances of M fall into three groups: those that are realized by P_1, those realized by P_2, and those realized by P_3. Call them M_1, M_2, and M_3 respectively. As an example: if M is the event–kind pain, M_1 may be human pains, M_2 reptilian pains, and M_3 Martian pains. Given that each instance of M_i inherits its causal powers from P_i, then, M_1, M_2, and M_3 as a group are going to be exactly as causally heterogeneous as P_1, P_2, and P_3, and M, as the union (disjunction, if you wish) of the M's, is going to be just as causally heterogeneous and lacking in nomic unity as the disjunction $P_1 \vee P_2 \vee P_3$. The conclusion seems inescapable: M fails to be a nomic property for the same reason that disqualifies $P_1 \vee P_2 \vee P_3$ as nomic.

What is the philosophical moral of all this? I will try to draw a broad picture of the general philosophical position I find attractive, stating its main claims without detailed explanation or supporting arguments. First of all, we lose mental properties as nomic properties —that is, as significant scientific properties. If they are multiply realizable in indefinitely many physical structures which can be extremely diverse from each other, they cannot show the kind of causal–nomological systematicity required of scientifically fruitful properties. If you are skeptical about properties purportedly expressed by heterogeneous disjunctive predicates, you should extend your skepticism to mental properties as well. Notice that my remarks are intended to apply only to mental properties; my arguments here say nothing about mental *concepts* or *predicates*: it may well be that mental concepts are not definable in terms of physical–biological concepts, and functionalism may well be a correct theory about meanings of mental expressions.[18] It's only that no properties answer to them —or, if you wish, only "wildly disjunctive" properties answer to them.

Terms for artifacts, e.g., "table", "carburetor" and the like, are often cited by functionalists to illustrate the causal–functional aspect of mental terms: anything that can serve a certain specific causal role, regardless of how it is physically

[18] Note that functionalism as it was originally formulated by Putnam in "Psychological Predicates" was a thesis about psychological concepts, or the meanings of psychological predicates (as is indicated by the title of the paper). In that article he sharply distinguishes concepts (or meanings) from properties.

realized, counts as a carburetor, or a table. But the analogy, if it is a good one, may not end there: just as we do not expect scientific laws and theories formulated in terms of "table", "carburetor", and the like, we should not expect laws and theories formulated in terms of mental expressions. This extension of the analogy seems unavoidable since the possibility of heterogeneous physical realization appears to the basic premise in both cases. This of course does not mean that there cannot be rough and ready, limited but useful, generalizations and rules of thumb about tables and carburetors, or about pains and mental images; nor does it mean that there aren't tables or carburetors, or pains. It only means that *being a table*, or *being in pain*, may well not be a nomic property fruitful for scientific theorizing.

Second, concerning psychophysical reduction: in a sense, there are no mental properties for which to search for uniform physical reductions. (Mental concepts or predicates have already been conceded to be physically irreducible —that is, not definable in biological or microphysical terms.) If you insist that there are mental properties whose reducibility we ought to discuss, the reply is that you can suit yourself by reducing them to the disjunctions of their physical realizers. The heterogeneity of the latter is only matched by the heterogeneity of the former; there is no cause to complain about the disjunctiveness of the reduction base.

So the overall metaphysical picture I recommend is this: mental properties are gone, at least as nomic properties of interest to scientific theory; mental concepts remain. Mental concepts may well be indefinable, for the familiar reasons; mental properties are not worth reducing, although, if anyone is interested, they can be considered reducible to disjunctions of their physical realizers. How does mental causation fare in this picture?

Suppose that an instance of M occurs in virtue of the fact that a physical property, P_1, realizes it. An earlier principle we have made use of says that the causal powers of this M–instance are exactly those of P_1. Why is this? Because there is no further fact of the matter about M's instantiation on this occasion over and beyond the fact that P_1 is instantiated. P_1's being instantiated, in an appropriate setting, simply counts as an instantiation of M.[19] Putting it this way is somewhat misleading, because it gives the impression that M is a genuine property; it would be more correct to say that P_1's being instantiated, in an appropriate setting, is correctly described by the use of the concept M (or predicate "M"). To speak of the causal powers of M (as such) is to speak disjunctively of the causal powers of each of its realizers. What causal contribution M makes in a given situation depends on what physical property realizes M on that occasion: M makes exactly the causal contribution made by its realizer. Having mental property M is causally efficacious because having M is either having P_1 or having P_2 or having P_3, and each of these is causally efficacious in its own way. (Note this important point: we need not say that having M is having the disjunctive property $P_1 \vee P_2 \vee P_3$.) So the causal powers of M have been reduced, multiply and disjunctively, to the causal powers of its physical realizers. This is all the reduction we need and can want, and the only kind of reduction appropriate for multiply realized properties.

[19] This is one point where the use of the "supervenience" idiom, rather than the "realization" idiom, can make a difference. If you believe in the supervenience of mental on physical properties but reject the doctrine that mental properties are realized by physical properties, you may very well hold that the instantiation of a mental property is a fact that is additional to, though supervenient on, the instantiation of an appropriate subvenient property.

9

PHILOSOPHICAL ISSUES, 3
Science and Knowledge, 1993

The Truth of Modest Realism

Ernest Sosa

Central concepts of epistemology —knowledge, for example, or epistemic justification, or intellectual virtue— derive in one way or another from that of truth. Assigning such a central role to truth makes one an externalist in epistemology. What is more, my externalism has a realist cast: the River Nile, the city of Khartoum, and the Mediterranean Sea are so related that the first flows from the second to the third, and does so independently of all beliefs, conventions, cultures, or conceptual schemes. Given only its meaning, the truth of my utterance about the Nile derives from a large and mind–independent feature of African geography. Moreover, when explorers and geographers, when astronauts and astronomers ply their trades, what drives them at least sometimes, at least to some degree, is just plain curiosity, the sheer desire to discover the truth in a certain realm, to find out the lay of the land, or of the heavens.

Such a perspective on knowledge and reality comes quickly into conflict with ideas on truth currently on the market. Richard Rorty's new pragmatism, for example, embraces a noncognitivist account of truth, justification, and knowledge.

Thus his neo–pragmatist

> ...thinks that the very flexibility of the word "true"...the fact that it is merely an expression of commendation ... ensures its univocity.... So he feels free to use the term "true" as a general term of commendation in the same way as his realist opponent does ... and in particular to use it to commend his own view.[1]

For the new pragmatism there is really no such thing as truth, nor any such thing as justification, nor anything like knowledge, any more than there is in the world any such thing as hurrayness. The term 'Hurray' is a functional term which serves for commending, and the like; for the "pragmatist", the terms 'true', 'justified', and 'known' are just like 'Hurray' in that respect. Thus:

> For the pragmatist ... "knowledge" is, like "truth", simply a compliment paid to the beliefs which we think so well justified that, for the moment, further justification is not needed.[2]

However, our epistemic vocabulary seems much more versatile than it could be were it merely a vocabulary of commendation such as the term 'Hurray'. Epistemic terms fit comfortably in many linguistic contexts where terms of commendation such as 'Hurray' would be out of place.[3]

A second well known approach explicates a true belief as one that would be accepted rationally and justifiably by ideal inquirers in an ideal epistemic situation regarding that belief

---

[1] "Solidarity or Objectivity?" in *Post–Analytic Philosophy*, ed. by J. Rajchman and C. West (New York: Columbia University Press, 1985), p. 6.

[2] *Ibid.*, p. 7.

[3] The problem was first raised by Peter Geach in "Ascriptivism", *Philosophical Review* 69 (1960): 121-5; and then in "Assertion", *Philosophical Review* 74 (1965): 449-65. Rortean noncognitivism and other philosophical views of truth are discussed more fully in my "Epistemology, Realism, and Truth", in the forthcoming volume on philosophy of language of *Philosophical Perspectives*(1993).

and its subject matter.[4] This Peircean approach has been much discussed and criticized, and I have myself objected to it elsewhere. Here I just want to note that if we wish to account philosophically for what is rational and epistemically justified by reference to the true, then we cannot concurrently accept a philosophical account of the true in terms of the rational and epistemically justified. So epistemic accounts of truth stay on the shelf as we move on.

A third sort of account of truth is correspondence to the facts, or correspondence with reality. For this view the sentences on our lips or in our brains are true when and only when they correspond appropriately with the facts or with reality. Such correspondence of a sentence with reality depends of course on the reference of its parts. But no naturalist can accept such aboutness, intentionality, or reference as a fundamental relation on a par with, let us say, spatiotemporal relations. So there would have to be some naturalisticaly acceptable account of such reference. And it is hard to imagine what form such an account might take except that of a causal account, as developed in diverse ways by Saul Kripke, Hartry Field, Hilary Putnam, Dennis Stampe, Fred Dretske, Michael Devitt, Ned Block, and others.[5] An outline of the view, applied to a simple case, might go like this. The sentence 'Planet Earth is round' is true in virtue of the fact that 'Planet Earth' refers to Planet Earth, 'is round'

---

[4] The history of epistemic truth would highlight Charles S. Peirce, "The Fixation of Belief", *Popular Science Monthly* 12 (1877): 1-15; Michael Dummett, *Truth and Other Enigmas* (Cambridge, MA: Harvard University Press, 1978); and Hilary Putnam, *Reason, Truth, and History* (Cambridge, UK: Cambridge University Press, 1981).

[5] S. Kripke, "Naming and Necessity", in *Semantics of Natural Language*, ed. D. Davidson and G. Harman (Dordrecht: Reidel, 1972); H. Field, "Tarski's Theory of Truth", *Journal of Philosophy* 69 (1972): 347-75; H. Putnam, "The Meaning of Meaning", in K. Gunderson, ed., *Language, Mind, and Knowledge: Minnesota Studies in the Philosophy of Science* 7 (Minneapolis: University of Minnesota Press, 1975); D. Stampe, "Toward a Causal Theory of Linguistic Representation", *Midwest Studies* 2 (1977); F. Dretske, *Knowledge and the Flow of Information* (Cambridge, MA: MIT Press, 1981); M. Devitt, *Designation* (New York: Columbia University Press, 1981); N. Block, "Advertisement for a Semantics for Psychology", *Midwest Studies* 10 (1986).

"refers" to roundness, and Planet Earth has roundness; and each of 'Planet Earth' and 'is round' refers to its referent in virtue of some appropriate causal connection that provides the required linkage.

Interesting consequences have recently been drawn from such a correspondence view by Stephen Stich,[6] who repudiates the intrinsic value of epistemic justification, and repudiates even its extrinsic value, even on a reliabilist conception of epistemic justification which would make "epistemically justified" belief likely to be true. For he extends his iconoclasm to truth itself, concluding that true belief should be granted neither intrinsic nor extrinsic value. Stich shows that philosophy can be fun, and delivers with great gusto and *éclat.* But what do we find the morning after, in a more sober reading?

Stich argues as follows:

a. True belief is not intrinsically valuable.

b. It is far from obvious that true belief is extrinsically valuable.

c. True belief is (therefore) not valuable; or at least it is far from obvious that it is valuable.

And how does he argue for his first premise, premise a? First we are told that the determination of what brain sentences count as (correspondence–)true is highly "partial" and "idiosyncratic", for it depends on what refers to what, and this turns out to depend on such partial and idiosyncratic determination. Indeed, what counts as referring to what is said to be significantly dependent on culture and tradition, like rules of etiquette. And if we aim only for such correspondence truth, we are likely to miss other head–world relations with better payoff.[7] In view of this, but in any case, we do

---

[6] *The Fragmentation of Reason* (Cambridge, MA: MIT Press, 1990).

[7] Here are some relevant passages from *The Fragmentation of Reason*:

> [The]... causal/functional interpretation is not only limited but also idiosyncratic. It is one interpretation function among many that stands out among its fellows principally because it is the function favored by local, contemporary commonsense intuition and

well to be "...not much inclined to value what is traditional and familiar for its own sake in matters epistemic".[8] Finally, those reasons "...may well be reason enough to decide that ...[we] do not really value true beliefs intrinsically".[9]

As for the second premise, the argument is much simpler: namely, that it is far from obvious that true beliefs are extrinsically valuable inasmuch as it is far from obvious that they are instrumentally optimal.[10]

> the largely unknown psychological processes that underlie that intuition (p. 117).
>
> [The]...interpretation function has a very limited domain. What this means is that there is a huge space of possible systems of mental computation and storage whose component states have no truth conditions and thus cannot be true. No doubt many of the systems in this semantics–free space are useless and chaotic. But there is certainly no reason to suppose that they all are. A much more likely possibility is that in this huge space there are systems that would vastly increase their user's power or happiness or biological fitness, systems that would lead to substantial reductions in the amount of suffering in the universe...(p. 119).
>
> [In the]...bristling infinity of functions there is one that is singled out by common sense as providing the "real" truth conditions for mental states, in contrast with the TRUTH* CONDITIONS, the TRUTH** CONDITIONS, and all other variations on the theme. But if, as we have been assuming, it is the causal/functional interpretation function that is sanctioned by intuition, then it is not a particularly simple or natural function. Rather, it is something of a hodgepodge, built from a more or less heterogeneous family of strategies for transmitting reference.... Those who find intrinsic value in holding true beliefs (rather than TRUE* ones, or TRUE** ones,...) are accepting unreflectively the interpretation function that our culture (or our biology) has bequeathed to us and letting that function determine their basic epistemic value (pp. 119-120).

[8] Stich, *op. cit.*, p. 120.

[9] *Ibid.*

[10] A relevant passage is on p. 124:

> [The] ...instrumental value of true belief is far from obvious. It is surely not the case that having true beliefs is *always* the best doxastic stance in pursuing our goals, and it would be no

Compare the following argument:

d. Eating delicious food is not intrinsically valuable.

e. It is far from obvious that eating delicious food is extrinsically valuable.

f. Eating delicious food (therefore) is not valuable; or at least it is far from obvious that it is valuable.

And how might one argue for the first premise, premise d? One might point out the partiality and idiosyncracy of what counts for us as eating delicious food. Indeed, what counts as such must be significantly dependent on culture and tradition, like rules of etiquette. The effect of fads, fashions, peer pressure, etc., on food tastes is surely large, even if not so large as their effect on clothing. But if we aim to eat only food that we now find delicious we are likely to miss other mouth–to–world relations with better payoff. In view of this, but in any case, don't we do well to be "not much inclined to value what is traditional and familiar for its own sake" in matters gastronomic? And, finally, are not such reasons "reason enough to decide that we do not really intrinsically value" eating delicious food, i.e., food that we now find delicious?[11]

As for the second premise, the argument would be much simpler: namely, that it is far from obvious that eating delicious food is extrinsically valuable inasmuch as it is far from obvious that it is instrumentally *optimal* (among all possible systems of substance–ingestion).

---

easy matter to show that believing the truth is *generally* (or even occasionally!) instrumentally optimal.

Note the move from "not obviously optimal" to "not obviously valuable". The actual argument in the book designed to cast doubt on the instrumental or extrinsic value of true belief depends essentially on that move.

[11]The argument can be made, I believe, whether we interpret 'delicious food' as 'food that we would find delicious' (in the eating), or as 'food that we *do* find generally delicious' or as 'food that *I* (do or would) find delicious' or even as 'mooshi pork, e.g., food that we, or at least I, do or would, now or generally find delicious'.

These arguments against eating delicious food are designed to parallel in every significant respect the earlier arguments against harboring true beliefs, in such a way that neither set of arguments will be any stronger than the other. But the arguments against eating delicious food are weak, and, given the parallel, the arguments against harboring true beliefs seem no stronger.

I conclude that, despite its persuasive power, the reasoning proposed by Stich does not really unsuit correspondence truth as central concept of epistemology and as appropriate goal of inquiry.[12]

Nevertheless, correspondence truth still seems problematic. Its main problem at present is not an excess of unacceptable consequences, however, so much as a deficiency of acceptable content. If the truth of sentences —linguistic, mental, or cerebral— is to receive a substantive correspondence analysis, it is indeed plausible that the reference of sentential constituents will be crucially involved, and that such reference will work through some sort of appropriate causal relations. But much work is still required before we have an acceptable understanding of the causal relations involved, and a substantive correspondence theory still seems a distant objective. Accordingly, substantive correspondence truth also stays on the shelf, and we move on.

By this stage we can sense the attractions of a streamlined deflationism about the concept of truth. For deflationism, truth is not a property that needs a substantial theory such as the correspondence theory, or a coherence theory, or the traditional pragmatist theory that the true is what works (as opposed to what Rorty calls pragmatism, really a kind of noncognitivism), or the theory that the true is what is validated by one's culture, or any other such theory which tries to ascribe some cognitive content to the concept of truth, any

[12]In a longer treatment I would compare Stich's arguments against the value of truth and against the value of epistemically justified belief, given certain assumptions about the nature of these, with the arguments of Parfit that question the rationality of caring about one's own survival, given certain assumptions about the nature of personal identity through time. I consider closely the issues raised by Parfit in "Surviving Matters", *Nous* 24 (1990): 305-30.

such theory which considers truth a property whose "nature" needs to be explained. A recent advocate of deflationism, Michael Williams, puts it like this:

The disquotation schema tells us just about all we need to know about truth.[13]

> It is *not* the case that "...truth is the name of a property whose nature neither the disquotation schema nor remarks on the use of 'true' to commend etc. so much as hint at, but which theories of truth, properly so called, must try to explain".[14]

Why has it been supposed that we need any such substantive property of truth anyhow? Two main reasons are on offer, according to Williams, and in a way these are really two forms of the same reason. The first is the view that we need to appeal to truth in order to explain the success that we enjoy in negotiating an ordinary day. At every turn we depend on our beliefs, most of them quite banal and trivially true; fortunately so, since we stake a lot on them, including our survival. We succeed as well as we do partly because we are right as often as we are, because we have many true beliefs that serve us well given our interests and desires. Similarly for our collective success through the use of our technology. Here we succeed as well as we do in part because the scientific theories which underlie such technology are nearly enough true, often enough. To these arguments Williams responds as follows:

> To spell out the explanation [of the success of our methods by the truth of our theories]... we should have to assert, in a qualified way, all the theories we currently accept, or all those belonging to "mature" sciences. But the predicate 'true' saves us the trouble, for we can compress this tedious rehearsal of current views into a single sentence and say "Our methods work because the theories that inform them are approximately true".[15]

---

[13] M. Williams, "Do We (Epistemologists) Need a Theory of Truth", *Philosophical Topics* 14 (1986): 223.

[14] *Ibid.*

[15] *Ibid.*, p. 230.

> If I want a cold drink and believe that the refrigerator, rather than the oven, is the place to look, I will increase the likelihood of getting what I want. This is because cold drinks are more likely to be found in the refrigerator than in the oven. To say that my having true beliefs makes it more likely that I will attain my goals is just a compact way of pointing to an indefinite number of mundane facts of this sort.[16]

This is widely regarded as an adequate response to the claim that we need substantive truth in order to explain our individual or collective success, what degree of it we enjoy anyhow. I confess to some lingering doubt, myself, based on my uncertain grasp of the relevant operations of "abbreviating" or "compressing" or "compacting" appealed to.[17] *Stipulative* "abbreviation" is of course clear and unobjectionable, but, just as clearly, that is not all the "abbreviating" that — on the proposal under discussion— the vocabulary of truth must be expected to do. Even once this had been explained adequately, moreover, what would follow from the fact that a certain vocabulary enabled one to abbreviate in the way suggested? Would it establish that no real properties are expressed by such vocabulary? Is there no real property of the world expressed by "For all x, Fx" just because this may be viewed as some sort of abbreviation of "Fa_1, Fa_2,..."? There would be some irony in the result that universal quantification has no real place in scientific explanation, since it is just a way of abbreviating a long conjunction, perhaps an infinite conjunction. Well, if truth turns out to have in explanation a place analogous to that of universal quantification, perhaps that is good enough for any point in the making, whether we end up calling truth a real property or not. Perhaps its status is secure enough if it is no less shaky than that of quantification.

A deflationist is not unwilling to define truth, so long as the definition is not of any traditional, "substantive" va-

---

[16] *Ibid.*, p. 232.

[17] In what follows I will abbreviate 'abbreviating, compacting, or compressing' as just 'abbreviating'.

riety. Some deflationists accept, for example, a substitutionalist paraphrase.[18] Does the sentence "Her statement is true" attribute a "property" to her statement? This question reduces through the substitutionalist paraphrase to the question whether in the following we have such a property: "For all *p*, if *p* is identical to <u>her statement</u>, then *p*". Does the part of this sentence that is not underlined express a property attributed to the referent of the underlined singular term? Well, if it expresses *something*, whatever it expresses is true of her statement contingently, and *not* true of many other statements. So it expresses something that is contingently true of some entities and contingently false of other entities. It is this "property", moreover, that we appeal to when we explain our individual and collective success by saying that enough of our pertinent beliefs and theories have it. And it is presumably this property that we appeal to in supposing that every instance of the disquotation schema —*p* is true–in–our–language–*L* iff *p*— is "true", and obviously so. The supposed feature that every instance of the disquotation schema is thus said obviously to have is then presumably just the feature attributed to "her statement" above: that of being such that for all *p*, if *p* is identical to it, then *p*.

However, not all forms of deflationism are associated with substitutional quantification. An attractive alternative is offered by Paul Horwich, who explicitly rejects the substi-

[18] Cf. Williams, *op. cit.*, p. 234:

> The disquotation schema reveals truth as a device for semantic ascent and descent: in other words, it suggests that talk of truth can always be paraphrased away. An obvious way to eliminate talk of truth from... laws linking confirmation by our methods with truth would be to rewrite them as
>
> For any *p*, if *p* is confirmed by our methods then probably *p*.
>
> True, this involves quantifying over propositions or statements. But this is something we might want to do anyway —e.g., to handle sentences like "Everything he said was true".

tutional account, and whose view focuses on the following schema:[19]

1. The proposition that p is true iff p

Anyone who has the concept of truth, notes Horwich, anyone who grasps the property of truth, will accept all propositions expressed by instances of 1 with respect to particular declarative sentences of their idiolect. Horwich's minimalist theory MT is constituted by all such instances of the "equivalence schema" 1 above.: "Our theory of truth ... is a collection of propositions —those expressed by instances of... [1]".[20]

We are warned that the theory of truth is irremediably infinite, that "...we should not expect to contain all instances of the equivalence schema within a finite formulation: an infinity of axioms is needed. And since this would seem to be an unavoidable feature of any adequate theory of truth, it should not be held against MT. Therefore we must acknowledge that the theory of truth cannot be explicitly formulated. The best we can do is to give an implicit specification of its basic principles".[21]

In showing how MT can explain all the phenomena of truth predication that need explaining, the following is postulated: that given "...the meaning of 'implies', we have every proposition of the form $\langle\langle p\rangle\text{implies}\langle q\rangle \rightarrow (p \rightarrow q)\rangle$".[22] (Read: "The proposition that–p implies the proposition that–q *only if*: p *only if* q", reading the '$\rightarrow$' as the material conditional.) Hence, by the meaning of 'entails', we must also have the principle schema:

(PE) The proposition that–p entails the proposition that–q *only if*: p *only if* q.

And, given PE, why not formulate a finitely minimal theory of truth as follows:

---

[19] In his pithy and convincing *Truth* (Oxford: Basil Blackwell, 1990).
[20] *Ibid.*, pp. 37-8.
[21] *Ibid.*, p. 31. Cf. p. 21, and pp. 11-2.
[22] *Ibid.*, p. 23.

(FMT) Every proposition is necessarily equivalent to (entails and is entailed by) the proposition that it is true.[23]

This theory captures the infinity of axioms constitutive of MT and, obviously, does so through a finite formulation. Or so I will now argue.

We begin with a more symbolic formulation of FMT:

a. $(x)[x \Leftrightarrow \langle x \text{ is true}\rangle]$	FMT
b. $\langle$Snow is white$\rangle \Leftrightarrow \langle\langle$Snow is white$\rangle$ is true$\rangle$	a, Univ. Inst.
c. Snow is white *iff* $\langle$Snow is white$\rangle$ is true	b, PE

And, in general, take any arbitrary member of MT, expressed by some arbitrary instance of the equivalence schema E: $\langle p\rangle$ is true *iff* p. We can now derive it as follows:

a. $(x)[x \Leftrightarrow \langle x \text{ is true}\rangle]$	FMT
b. $\langle p\rangle \Leftrightarrow \langle\langle p \text{ is true}\rangle \text{ is true}\rangle$	a, Univ. Inst.
c. p *iff* $\langle p\rangle$ is true	b, PE
d. $\langle p\rangle$ is true *iff* p	c, Truth–functional Logic

Therefore, given only the familiar modal notion of entailment among propositions, we can after all capture the infinity of MT within the finite formulation provided by FMT. At least, we can do so provided we are allowed appeal to PE.

In any case, for Horwich minimalism applies not only to propositional truth but also to truth as a property of beliefs and even utterances. Thus he writes on p. 7: "...I shall go on to argue that the minimalist conception applies equally well to the 'truth' of utterances, mental attitudes, and other types of entity". And he goes on to formulate an interesting "minimalist" theory of truth for utterances, one that is meant

[23] 23 More explicitly, FMT reads: "Every proposition *P* is necessarily equivalent to (entails and is entailed by) the *Russellian* proposition that it is true: the *de re*, "concrete" proposition, with respect to *P*, of its being true". For an account of such "Russellian" propositions, see *Truth*, pp. 94-5.

to deal with problems raised by indexicals, demonstratives, and ambiguity. But actually it seems to me that we can best isolate some main philosophical issues if we focus on his first approximation to a minimalism of truth for utterances, as it appears on p. 103: "The initial deflationary impulse is to say that

(D?) Any declarative utterance of the sentence–type '*p*', is true iff *p*."

Let us restrict ourselves to sentences that have no indexicals or demonstratives and that are perfectly univocal: call them "eternal sentences". Is (D?) an acceptable minimalism of truth for utterances of eternal sentences?

A problem with (D?) is that it does not give us a general set of necessary and sufficient conditions for the truth of an utterance, nor does it even, like MT, give us a theory in terms of all the instances of some biconditional schema. It only says that *if u* is a declarative utterance of a sentence–type '*p*', *then u* is true iff *p*. We can overcome this with a two–part theory as follows:

D_1 Utterance *u* is true iff *u* is an utterance of *a* (translation of *a*) sentence type '*p*' that is true.

D_2 The sentence *Snow is white* is true iff snow is white, the sentence *The sky is blue* is true iff the sky is blue,. . . , and so on, for all the instances of the equivalence schema:

(E–sent) The sentence *p* is true iff *p*

obtained by filling in with declarative sentences of English or possible extensions of English.

Let 'D' stand for the "conjunction" of D_1 and D_2. A problem with D is that it is no longer "pure". It involves a relation between an utterance and a sentence type of that utterance being *an utterance of* (a translation of) that sentence type. And this is not just a simple relation of the utterance being of a certain shape, or the like. It is not just a syntactic matter. On the contrary, an utterance in German of 'Schnee ist weiss' would fit D by being an utterance of (a translation of) 'Snow is white', and what makes it so is the fact that it is an utterance of 'Schnee ist weiss' while 'Schnee ist weiss' is translatable as 'Snow is white'. And with this latter relation we get into the matter of linguistic use or role of an expression in a linguistic community as its members interact with each other and with the surrounding world. So it is hard to see how D can be in any appropriate sense "minimalistic." It seems very unlike MT in this respect.

Using a similar line of reasoning to that which leads Horwich to his minimalist theory of truth MT, we may supplement that theory of truth, constituted by such an infinity of instances of 1, with a similar theory of *expression*, as follows. Consider the schema

2. The sentence *p* expresses the proposition that *p*

where the context indicates that in any instance of 2 the sentence that replaces the schematic letter '*p*' will in its first occurrence function autonymously. Anyone who has the concept of expression, who grasps that semantical relation will accept all propositions expressed by instances of 2 with respect to particular declarative sentences of their idiolect.

Assume next:

3. X is a fact $=_{\mathrm{Df}}$ X is a true proposition

4. (Sentence) S corresponds to a fact $=_{\mathrm{Df}}$ S expresses a fact

5. (Sentence) S is true $=_{\mathrm{Df}}$ S expresses a true proposition

It follows that:

6. (Sentence) S is true $\equiv$ S corresponds to a fact.

And with 6 we arrive at a sort of correspondence account of the truth of sentences. However, this obviously is not in competition with the aim of developing a more substantive correspondence account in terms of the reference of sentence constitutents. Within our present framework, that aim would focus more specifically on a theory of *expression*, a theory of how a sentence expresses a proposition in virtue of its parts referring to whatever it is that they refer to. It is true, of course, that we can give a similarly abstract account of the reference of our terms, using principles such as:

The predicate F expresses the property of being F

and the propositions expressed by instances of this schema with respect to particular predicates of our idiolect. This still would leave open the question of what such predicate/property pairs have in common, however, of what it is in virtue of which predicates express the properties they express. It is this further question that the more substantive correspondence theory tries to answer. But what if there *is* no answer, or at least none that *we* can find? Does that undermine any epistemology that gives truth pride of place?

Not at all. It now seems likely that Moore and Russell were right in their early, too soon discarded, conception of truth, a conception described by Moore much later as follows:

> It is a theory which I myself formerly held, and which certainly has the advantage that it is very simple. It is simply this. It adopts the supposition that in the case of every belief, true or false, there is a proposition which is what is believed, and which certainly is. But the difference between a true and a false belief, it says, consists simply in this, that where the belief is true the proposition, which is believed, besides the fact that it *is* or "has being" also has another simple unanalysable property which is possessed by some propositions and not by others. The propositions which don't possess it, and which therefore we call false, *are* or "have being" —just as much as those which *do*; only they just have *not* got this additional property of being "true".[24]

N.B.: What this denies is that truth has a Moorean "analysis" or definition.[25] It does not deny that we may know a

[24] From p. 261 of the published version of Moore's 1910-11 lectures, *Some Main Problems of Philosophy* (London: Allen and Unwin: 1953). For discussion of the Moore/Russell view, see Richard Cartwright's "A Neglected Theory of Truth", in his *Philosophical Essays* (Cambridge, MA: MIT Press, 1987), pp. 71-95.

[25] Unless we countenance a substitutional account, which would not be a substantive account as traditionally understood.

priori lots of propositions constituted in part by our primitive concept of truth.

When Moore viewed truth as a simple, unanalyzable quality, he viewed it just the way he viewed good in *Principia Ethica*. So he surely would have said of truth what he did say of good as follows:

> If I am asked "What is good?" my answer is that good is good, and that is the end of the matter. Or if I am asked "How is good to be defined?" my answer is that it cannot be defined, and that is all I have to say about it.... My point is that "good" is a simple notion, just as "yellow" is a simple notion; that, just as you cannot, by any manner of means, explain to anyone who does not already know it, what yellow is, so you cannot explain what good is. Definitions...which describe the real nature of the object or notion denoted by a word ...are only possible when the object or notion in question is something complex.... But yellow and good, we say, are not complex: they are notions of that simple kind, out of which definitions are composed and with which the power of further defining ceases.[26]

On this view, what you cannot do either with good or with yellow or with truth is to define it, to give an illuminating, compact, at least surveyable, Moorean analysis of it. It is in *this* sense that you cannot philosophically "explain" any such "simple" concept. And this leaves it open that you should know *a priori* propositions, perhaps infinitely many, constituted essentially by such concepts, propositions perhaps like the following:

> No surface that is yellow all over is blue all over.
>
> Pleasure is intrinsically good.
>
> That–snow–is–white is true iff snow is white.

The conclusions we have reached constitute a modest realism, MR, that accepts TE, truth–centered epistemology

[26] *Principia Ethica* (Cambridge, UK: Cambridge University Press, 1903), pp. 6-8.

(which views knowledge as belief acquired through truth-conducive faculties or abilities or intellectual virtues); and accepts OR, ontological realism (the view that reality, including any evaluative or mental reality, is as it is independently of any minding of it by any finite subjects or collections thereof). Note well: this does not deny the possibility of an in itself mental reality; all it requires is that this reality be as it is independently of any minding of *it*, i.e., independently of any attitudes *distinct* from that mental reality itself, while giving rise to it and constituting it. And it is crucial here to distinguish the sort of independence at issue. For it may even be granted by MR that one could not possibly be certain that one has a headache without having it, and, more generally, that one could not possibly have ideal warranted or apt or epistemically justified belief without that belief being true. None of this would show that the object of belief, the headache or whatever, is in the relevant sense *dependent* on the believing. True, the believing could not in those circumstances be there without the object of belief being there. But that is not sufficient to show that the object of belief is in the relevant sense dependent on the believing. It would not show that the believing "gives rise" to or "constitutes" its object. (Compare here the Euthyphro controversy as to whether the good is good in virtue of being the object of the gods' love or whether on the contrary it is the object of the gods' love in virtue of being good.) Anyhow, concerning such modest realism, MR, we have reached the following conclusions.

1. Modest realism (MR) can and should reject epistemic accounts of truth as constituted by what would be believed by ideal inquirers in ideal conditions (despite the case made by Peirce and Putnam in favor of such an account), as well as any oxymoronically noncognitivist neopragmatism of knowledge and truth, according to which our vocabulary of truth and knowledge has only a commendatory function (despite Rorty's writings to that effect). And MR can retain a view of truth as the proper goal of inquiry and as an important epistemic value (despite the recent arguments by Stich).

2. MR need not take a position on how our thoughts gain their contents, on how they are enabled to refer or correspond to an independent or objective reality. (It may be argued, however, that MR, because of its commitment to ontological realism (OR) is precluded from ever giving an acceptable account of reference or intentionality or correspondence, since any such account would need to appeal to a relation of causation that could not be absolute or objective but must be on the contrary perspectival and context-dependent. But this objection may be answered by reference to 3 and 4 below.)

3. MR can accept a notion of correspondence or reference according to which a word or a brain state of ours can refer to some external or objective or independent reality not objectively or absolutely but only perspectivally or contextually: in the sense that '*W* refers to *O*' is not true or false period, but may be true relative to one context or perspective while it is false relative to another. (This no more forces us to reject OR than does the fact that 'I now sit' is not true or false period, even if no satisfactory view of ourselves in the world can entirely exclude such perspectival indexicals as 'I' and 'now' or the corresponding indexical concepts.)

4. But anyhow we need not opt even for a perspectival correspondence account of truth, since we can rest with Horwich's minimalist theory MT that simply accepts every instance of 'The proposition that-*p* is true iff *p*'. What is more, we might even opt for a finitely minimal theory of truth, FMT, whose basic principle is simply: Every proposition is necessarily equivalent to the (Russellian) proposition that it is true. This turns out to capture the infinity of MT through a finite formulation, at the price of using the modal concept of necessary equivalence.

5. None of this amounts to a *definition* of truth or to any meaning analysis of it. So in a clear sense we are accepting the "primitivism" of truth defended by Moore and Russell for a time. Truth is a primitive concept that

has no explicit definition even though it has a place in simple *a priori* propositions such as those constitutive of MT, and in simple *a priori* general principles such as FMT.

6. Primitivism is thus defensible with regard to the concept of truth applicable to propositions; what is more, it seems unavoidable with regard to the truth that applies to utterances, sentences, and beliefs. Accepting this is only to accept that we shall never have a simple explicit definition or meaning analysis of the concept of truth that applies to such entities. But this is not to reject the possibility of our ever having a philosophical theory of such truth. Of course, it may be that under philosophical scrutiny such truth turns out to be best understood as a perspectival concept. But even this can be regarded by MR as an open possibility.

10

PHILOSOPHICAL ISSUES, 3
Science and Knowledge, 1993

Take Back the Day! Jon Dorling's Bayesian Solution of the Duhem Problem

Richard Jeffrey

> "It's all up for grabs; methodology is nothing but politics" (or sociology, or literary criticism).

Jon Dorling [1, 3] effectively rebuts that skeptical conclusion, a conclusion often drawn from the presumed arbitrariness of answers to the question of which to give up —theory or auxiliary hypothesis?— when they jointly contradict empirical data. As drawn by some of Quine's readers, the conclusion depends on the assumption that aside from our genetical and traditional heritage (including our sense of simplicity), deductive logic is all we've got to go on when it comes to theory testing. This leaves things pretty much as Descartes saw them, just before the mid–17th century emergence in the hands of Fermat, Pascal, Huygens and others of the "Bayesian" methodology that Dorling brings to bear on various episodes in the history of science.

The skeptical conclusion is one that scientists themselves generally reject, thinking they have good reason to evaluate the effects of evidence as they do, but regarding formulation and justification of such reasons as someone else's job —the methodologist's. Dorling addressed the Duhem problem 12 years ago in Bayesian terms,offering tongue–in–cheek support [1] for Imre Lakatos's views about the hard core and protective belt. (See [4], pp. 96-102, for a brief account.) Michael Redhead's elaboration [2] of Dorling's solution appeared in the following year. Here is an introduction to some of Dorling's subsequent progress [3] on the job.

I'll present it in terms of "Bayes factors":

The Bayes factor for T against S is the factor $f(T,S)$ by which the prior odds $p(T) : p(S)$ must be multiplied to obtain the posterior odds, $p(T|D) : p(S|D)$, in the light of empirical data D.

By Bayes' theorem,

$$\begin{array}{ccccc} & & \text{Bayes factor} & = & \text{Likelihood ratio} \\ f(T,S) & \stackrel{\text{def}}{=} & \dfrac{p(T|D)/p(S|D)}{p(T)/p(S)} & = & \dfrac{p(D|T)}{p(D|S)} \end{array}$$

The empirical result D is not generally deducible or refutable by T alone, or by S alone, but in interesting cases of scientific hypothesis testing D is deducible or refutable on the basis of the theory and an auxiliary hypothesis H (e.g., the hypothesis that the equipment is in good working order). To simplify the analysis, Dorling makes an assumption that can generally be justified by appropriate formulation of the auxiliary hypothesis:

Prior independence: $p(HT) = p(H)p(T)$, $p(HS) = p(H)p(S)$

In some cases S is simply the denial, $-T$, of T; in others it is a definite scientific theory R, a rival to T. In any case Dorling expands the right–hand side of the equation *Bayes Factor = Likelihood Ratio* using the independence assumption:

$$f(T,S) = \frac{p(D|TH)p(H) + p(D|T-H)p(-H)}{p(D|SH)p(H) + p(D|S-H)p(-H)} \qquad (1)$$

To study the effect of D on H, he also expands f(H,-H) with respect to T

$$f(H,-H) = \frac{p(D|HT)p(T) + p(D|H-T)p(-T)}{p(D|-HT)p(T) + p(D|-H-T)p(-T)} \quad (2)$$

and, similarly, with respect to S.

With observational data D and auxiliary hypothesis H, the pre–skeptical hypothetico–deductive terminology is:

Confirmation of T	$p(D\|HT) = 1$
Disconfirmation of R	$p(D\|HR) = 0$
Crucial test, for T against R	$p(D\|HT) = 1, p(D\|HR) = 0$

with probabilities 1 and 0 since HT implies D and HR implies $-D$. Dorling dissects out various conditions under which that terminology is appropriate or deceptive.

Thus, in a crucial test, on the assumption that H is false, T and R may be expected to be equally unrevealing about D:

$$p(D|T-H) = p(D|R-H) = k. \quad (3)$$

If a test satisfying this condition confirms T against R, the Bayes factor will be

$$\begin{aligned} f(T,R) &= \frac{p(H) + p(D|T-H)p(-H)}{p(D|R-H)p(-H)} \\ &= \frac{p(H)}{k_p(-H)} + 1 > \frac{p(H)}{p(-H)} \end{aligned} \quad (4)$$

Then such a test magnifies the odds between T and R by a factor exceeding the prior odds on H, for by definition of $f(T,R)$ we have

$$\frac{p(T|D)}{p(R|D)} = f(T,R)\frac{p(T)}{p(R)}$$

(Note: this magnification may represent a decrease in R's probability rather than an increase in T's.)

In [3], sec. 4, Dorling analyzes two famous crucial experiments that were duplicated, with apparatus differing in seemingly unimportant ways, with conflicting results: one of the duplicates confirmed T against R, the other confirmed R against T. But in each case the scientific experts took the experiments to clearly confirm one of the rivals against the other. Dorling explains why the experts were right.

1 Case 1. Einstein vs. Newton, 1919

"In the solar eclipse experiments of 1919, the telescopic observations were made in two locations, but only in one location was the weather good enough to obtain easily interpretable results. Here, at Sobral, there were two telescopes: one, the one we hear about, confirmed Einstein; the other, in fact the slightly larger one, confirmed Newton. Conclusion: Einstein was vindicated, and the results with the larger telescope were rejected" ([3], p. 19).

Notation. T: General Relativistic light–bending effect of the sun; R: No light–bending effect of the sun; H: Both telescopes are working correctly; D: The actual, conflicting data from both telescopes

As $p(D|TH) = p(D|RH) = 0$, the Bayes factor is simply

$$f(T,R) = \frac{p(D|TH)p(H) + p(D|T-H)p(-H)}{p(D|RH)p(H) + p(D|R-H)p(-H)} = \frac{p(D|T-H)}{p(D|R-H)} \qquad (5)$$

"Now the experimenters argued that one way in which H might easily be false was if the mirror of one or the other of the telescopes had distorted in the heat, and this was much more likely to have happened with the larger mirror belonging to the telescope which confirmed R than with the smaller mirror belonging to the telescope which confirmed T. Now the effect of mirror distortion of the kind envisaged would be to shift the recorded images of the stars from the positions predicted by T to or beyond those predicted by R. Hence

$p(D|T - H)$ was regarded as having an appreciable value, while, since it was very hard to think of any similar effect which could have shifted the positions of the stars in the other telescope from those predicted by R to those predicted by T, $p(D|R-H)$ was regarded as negligibly small, hence the result as overall a decisive confirmation of T and refutation of R" ([3], p. 19).

2 Case 2. Bell's Inequalities: Holt vs. Clauser

"Holt's experiments were conducted first and confirmed the predictions of the local hidden variable theories and refuted those of the quantum theory. Clauser examined Holt's apparatus and could find nothing wrong with it, and obtained the same results as Holt with Holt's apparatus. Holt refrained from publishing his results, but Clauser published his, and they were rightly taken as excellent evidence for the quantum theory and against hidden–variable theories" ([3], p. 20).

Notation. T: Quantum theory; R: Disjunction of local hidden variable theories; H: Holt's setup is sensitive enough to distinguish T from R; D: The specific correlations predicted by T and contradicted by R are not detected by Holt's setup.

The characterization of D yields the first two of the following equations. In conjunction with the characterization of H it also yields $p(D|T - H) = 1$, for if H is false, Holt's apparatus was not sensitive enough to detect the correlations that would have been present according to T; and it yields $p(D|R - H) = 1$ because of the wild improbability of the apparatus "hallucinating" those specific correlations.

$$p(D|TH) = 0,\ p(D|RH) = 1,\ p(D|T-H) = p(D|R-H) = 1$$

Making these substitutions, we have

$$f(T, R) = \frac{p(D|TH)p(H) + p(D|T - H)p(-H)}{p(D|RH)p(H) + p(D|R - H)p(-H)} = p(-H) \tag{6}$$

Then with a prior probability 4/5 for adequate sensitivity of Holt's apparatus, the odds between quantum theory and the local hidden variable theories shift strongly in favor of the latter, e.g., with prior odds 45:55 between T and R, the posterior odds are only 9:55, a 14% probability for T.

Why then does not Holt publish his result? Because the experimental result undermined confidence in his apparatus. Setting $-T = R$ in (2) because T and R were the only theories given any credence as explanations of the results, and making the same substitutions as in (5), we have

$$f(H,-H) = \frac{p(D|HT)p(T) + p(D|HR)p(R)}{p(D|-HT)p(T) + p(D|-HR)p(R)} = p(R) \tag{7}$$

so that the odds on H fall from 4:1 to 2.2:1, the probability of H falls from 80% to 69%. Holt is not prepared to publish with better than a 30% chance that his apparatus could have missed actual quantum mechanical correlations; the swing to R depends too much on a prior confidence in the experimental setup that is undermined by the same thing that caused the swing.

Now why did Clauser publish?

Notation. T: Quantum theory; R: Disjunction of local hidden variable theories; C: Clauser's setup is sensitive enough; E: The specific correlations predicted by T and contradicted by R are detected by Clauser's setup.

Suppose that $p(C) = .5$: at this point both experimenters trust Holt's well–tried set–up better than Clauser's. Suppose Clauser's initial results E indicate presence of the quantum mechanical correlations pretty strongly, but still with a 1% chance of error. Then E strongly favors T over R:

$$\begin{aligned} f(T,R) &= \frac{p(E|TC)p(C) + p(E|T-C)p(-C)}{p(E|RC)p(C) + p(E|R-C)p(-C)} \\ &= \frac{.5 + .01 \times .5}{.01} = 50.5 \end{aligned} \tag{8}$$

Starting from the low 9:55 to which T's odds fell after Holt's experiment, odds after Clauser's experiment will be 909:110, an 89% probability for T.

The result E boosts confidence in Clauser's apparatus by a factor of

$$f(C,-C) = \frac{p(E|CT)p(T) + p(E|CR)p(R)}{p(E|-CT)p(T) + p(E|-CR)p(R)} = 15 \quad (9)$$

This raises the initially even odds on C to 15:1, raises the probability from 50% to 94%, and drops the 50% probability of the effect's being due to chance down to 6 or 7 percent.

Finally, note one more class of cases: a theory T remains highly probable although (with auxiliary hypothesis H) it implies a false prediction D. With $S = -T$ in formulas (1) and (2), with $p(D|TH) = 0$, and setting

$$t = \frac{p(D|T-H)}{p(D|-T-H)} \qquad h = \frac{p(D|-TH)}{p(D|-T-H)}$$

we have

$$\begin{aligned} f(T,-T) &= \frac{0 + p(D|T-H)p(-H)}{p(D|-TH)p(H) + p(D|-T-H)p(-H)} \\ &= \frac{t}{h\frac{p(H)}{p(-H)} + 1} \quad (10) \\ f(H,-H) &= \frac{0 + p(D|H-T)p(-T)}{p(D|-HT)p(T) + p(D|-H-T)p(-T)} \\ &= \frac{h}{t\frac{p(T)}{p(-T)} + 1} \quad (11) \\ f(T,-H) &= \frac{0 + p(D|T-H)p(-H)}{0 + p(D|H-T)p(-T)} \\ &= \frac{t}{h} \times \frac{p(-H)}{p(-T)} \quad (12) \end{aligned}$$

Note how these formulas apply to two historical cases.

3 Case 3. Laplace vs. Adams: Secular Acceleration of the Moon

"Dorling (1979) gave a probabilistic analysis of a famous episode from the history of astronomy which clearly illustrated striking asymmetries in 'normal' scientists' reactions to confirmation and refutation. This particular historical case furnished an almost perfect controlled experiment from a philosophical point of view, because owing to a mathematical error of Laplace, later corrected by Adams, the same observational data were first seen by scientists as confirmatory and later as disconfirmatory of the orthodox theory. Yet their reactions were strikingly asymmetric: what was initially seen as a great triumph and of striking evidential weight in favour of the Newtonian theory, was later, when it had to be re–analyzed as disconfirmatory after the discovery of Laplace's mathematical oversight, viewed merely as a minor embarrassment and of negligible evidential weight against the Newtonian theory. Scientists reacted in the 'refutation' situation by making a hidden auxiliary hypothesis, which had previously been considered plausible, bear the brunt of the refutation, or, if you like, by introducing that hypothesis's negation as an apparently ad hoc face–saving auxiliary hypothesis" ([3], p. 1).

Notation. T: the theory, Newtonian celestial mechanics; H: The hypothesis that disturbances (tidal friction, etc.) make a negligible contribution to; D: the observed secular acceleration of the moon.

In [1,3] Dorling argues[1] in considerable scientific and historical detail for approximate numerical values

$$t = 1, \qquad h = 1/50$$

Then for a theorist whose odds are 3:2 on H and 9:1 on T (probabilities 60% for H and 90% for T),

$$f(T, -T) = 100/103, \quad f(H, -H) = 1/500, \quad f(T, H) = 200$$

[1] Summing it up very roughly: $t = 1$ because with H false, truth or falsity of T is irrelevant to D, and $t = 50h$ because in plausible partitions of $-T$ into rival theories predicting lunar accelerations, $p(R|-T) = 2\%$ where R is the disjunction of rivals not embarrased by D.

Thus the prior odds 900:100 on T barely decrease, to 900:103; the probability of T after the refutation, 900/1003, agrees with the original 90% to two decimal places. But odds on the auxiliary hypothesis H drop sharply from 3:2 to a posterior value of 3/1000, the probability of H drops from 60% to about three tenths of 1%.

4 Case 4. Prout and Thomson (Howson and Urbach, pp 96-102)

Here T is William Prout's (1815) and Thomas Thomson's (1818) hypothesis that the atomic weights of all elements are whole–number multiples of the atomic weight of hydrogen, D is the (1815) measurement of the atomic weight of chlorine as 35.83 times that of hydrogen, and H asserts accuracy of the measuring technique and purity of the measured samples, to ±.005, say. Estimating[2] $t = 2$, $h = 1$, $p(T) = 90\%$ and $p(H) = 60\%$, Howson and Urbach calculate $p(T|D) = 88\%$ and $p(H|D) = 7\%$. Thus, $f(T,H) = 8$.

5 Dorling's conclusions

"Until recently there was no adequate theory available of how scientists should change their beliefs in the light of evidence. Standard logic is obviously inadequate to solve this problem unless supplemented by an account of the logical relations between degrees of belief which fall short of certainty. Subjective probability theory provides such an account and is the simplest such account that we posess. When applied to the celebrated Duhem (or Duhem–Quine) problem and to

[2] Taking $p(D|-TH)$ = accuracy of the measurement (.01, say); $= p(D|-T-H)$ since with H and T false "one would not expect atomic weights to be biased towards any particular part of the interval between adjacent integers"; $= p(D|T-H)/2$ since with T true and H false "the probability of an atomic wight lying in any part of the 35-36 interval would not be distributed uniformly over the interval, but would be more concentrated around the whole numbers".

the related problem of the use of *ad hoc*, or supposedly *ad hoc*, hypotheses in science, it yields an elegant solution. This solution has all the properties which scientists and philosophers might hope for. It provides standards for the validity of informal deductive reasoning comparable to those which traditional logic has provided for the validity of informal deductive reasoning. These standards can be provided with a rationale and justification[3] quite independent of any appeal to the actual practice of scientists, or to the past success of such practices. Nevertheless they seem fully in line with the intuitions of scientists in simple cases and with the intuitions of the most experienced and most successful scientists in trickier and more complex cases. The Bayesian analysis indeed vindicates the rationality of experienced scientists' reactions in many cases where those reactions were superficially paradoxical and where the relevant scientists themselves must have puzzled over the correctness of their own intuitive reactions to the evidence. It is clear that in many such complex situations many less experienced commentators and critics have sometimes drawn incorrect conclusions and have mistakenly attributed the correct conclusions of the experts to scientific dogmatism. Recent philosophical and sociological commentators have sometimes generalized this mistaken reaction into a full-scale attack on the rationality of men of science, and as a result have mistakenly looked for purely sociological explanations for many changes in scientists' beliefs, or the absence of such changes, which were in fact, as we now see, rationally *de rigeur*.

"It appears that in the past even many experts have sometimes been misled in trickier reasoning situations of this kind. A more widespread understanding of the adequacy and power of the kinds of Bayesian analyses illustrated in this paper could prevent such mistakes in the future and could form a useful part of standard scientific education. It would be an exaggeration to say that it would offer a wholly new level of precision to informal scientific reasoning, for of course the quantitative subjective probability assignments in such calculations are merely representative surrogates for informal

[3] A long footnote defends Dutch book arguments to this effect.

qualitative judgments. Nevertheless the qualitative conclusions which can be extracted from these relatively arbitraty quantitative illustrations and calculations seem acceptably robust under the relevant latitudes in those quantitative assignments. Hence if we seek to avoid qualitative errors in our informal reasoning in such scientific contexts, such illustrative quantitative analyses are an exceptionally useful tool for ensuring this, as well as for making explicit the logical basis for those qualitative conclusions which follow correctly from our premises, but which are sometimes nevertheless surprising and superficially paradoxical" ([3], pp. 22-23).[4]

6 REFERENCES

[1] Jon Dorling, "Bayesian personalism, the methodology of research programmes, and Duhem's problem" Studies in History and Philosophy of Science 10(1979)177-187.

[2] Michael Redhead, "A Bayesian reconstruction of the methodology of scientific research programmes," Studies in History and Philosophy of Science 11(1980)341-347.

[3] Jon Dorling, "Further illustrations of the Bayesian solution of Duhem's problem," (29 pp., photocopied, undated: Centrale Interfaculteit, Universiteit van Amsterdam, Grimburgwal 10, Amsterdam).

[4] Colin Howson and Peter Urbach, Scientific Reasoning: the Bayesian approach, Open Court, La Salle, Illinois, 1988.

[4]I find the quantitative analysis of cases 1 and 2 more persuassive *prima facie* than that of cases 3 and 4; but the important question about cases 1 and 2 is whether scientists of the day would rightly be persuaded. Dorling [1, 3] and Howson and Urbach [4] offer detailde historical arguments for a positive answer to that question.

11

PHILOSOPHICAL ISSUES, 3
Science and Knowledge, 1993

What's to be said for simplicity?

Richard Foley

It is a commonplace that the topic of simplicity is anything but simple. This is especially so when the issue is whether or not considerations of simplicity give us a reason to believe a hypothesis. It is easy to get caught between the warring sides on this issue. On the one side are those who insist that considerations of simplicity are pervasive in all our theorizing and thus it had better be reasonable to make use of them. On the other side are those who cannot understand why anyone would think that the simplicity of a hypothesis is a reason to believe it —that is, a reason to believe that it is true.

I will defend a compromise position, although I suspect that the proponents of simplicity will see me as siding with the skeptics, since I will be suggesting that we are irrational to the degree that considerations of simplicity shape what we believe. But on the other hand, I will also be suggesting that the kind of irrationality at issue here may not be worth worrying about. I will even suggest that it may be rational for us to put up with this kind of irrationality in ourselves.

But this is the end of the story. The story begins more modestly, with our epistemic ends. Our epistemic ends are those concerned solely with the accuracy and comprehensiveness of our beliefs. It is in terms of these ends that epistemic reasons for belief are distinguished from non-epistemic reasons. Epistemic reasons "aim" at epistemic ends, and we are epistemically rational to the extent that we have adequate epistemic reasons for what we believe. Our evidence indicates, or at least purports to indicate, what propositions are likely to be true. In so doing it indicates what beliefs are likely to satisfy our epistemic ends, and it thus gives us epistemic reasons to have those beliefs.

There is little agreement about what makes something evidence for the truth of a proposition, and hence there is equally little agreement about what gives us an epistemic reason to believe a proposition. There are foundationalist accounts, coherentist accounts, and others as well. But what makes each of these an account of epistemic rationality is that it tries to describe what we have reasons to believe insofar as our aim is to have accurate and comprehensive beliefs.

Moreover, the aim is to have accurate and comprehensive beliefs now, not at some future time. After all, there can be situations in which believing an obviously false proposition will increase the likelihood of your having accurate and comprehensive beliefs at some future time. For example, let P be the proposition that the future intellectual development of humans depends on you alone, since you alone are a reliable inquirer. Believing *P* is likely to increase your intellectual dedication. As a result, it may also increase your long-term chances of having accurate and comprehensive beliefs. Even so, you don't have a reason to believe *P* insofar as your aim is now to have accurate and comprehensive beliefs. In other words, you don't have an epistemic reason to believe it.

This tie between epistemic reasons and current epistemic ends imposes a restriction upon accounts of epistemic rationality. Whatever one's proposed criterion of epistemic rationality, the criterion must be one that "aims" at these epistemic ends. This restriction has more bite than it might appear to have at first glance. Suppose an account of rational belief recommends inferences of kind *k*, and suppose in addition that making such inferences will increase over the long-run our chances for true beliefs. Even so, it may be that the account is not best interpreted as trying to describe when we have good epistemic reasons for believing something. To be so construed, it must be the case that the recommended inferences are likely not just to get us at the truth eventually but also to do so now.

Keeping this point in mind, consider the following rule of inference: All else being equal, believe the simplest hypothesis that explains all the data in the domain at issue. There are various ways in which simplicity can be understood, some of which would make the above rule border on the trivial, since they reduce simplicity to something that plausibly might be said to always give a reason. So, for example, if by definition an explanation is simple to the degree that its elements are not improbable relative to one another, then the above rule is relatively uncontroversial but also relatively uninteresting. In effect, it merely tells us to believe the most probable explanation when all else is equal.[1]

Or suppose that the simpler of two explanations is by definition the one that requires the fewest revisions in the other beliefs we have some reason to hold. Then once again, the recommendation that we prefer simpler explanations is relatively uninteresting, since in this belief-relative sense of simplicity our other beliefs always give us a reason to be suspicious of the less simple explanation.

On these reductive approaches, the epistemic work is being done by that which simplicity is supposedly reduced to. I say 'supposedly' because it is doubtful whether these approaches capture the common sense notion of simplicity. In any event, I will just stipulate that it is nothing of this sort that I have in mind. Instead, I will say that the simplicity of a hypothe-

---

[1] "...a hypothesis is complex rather than simple in this sense to the extent that it contains elements within it, some of which are unlikely to be true relative to others, thus making the hypothesis as a whole unlikely on a purely *a priori* basis to be true; it is simple to the extent that this is not the case". Laurence Bonjour, *The Structure of Empirical Knowledge* (Cambridge: Harvard University Press, 1985), 183. Bonjour is well aware that this is a minimalist conception of simplicity. Indeed, that's what makes it suited for his special purpose. His strategy is to use this seemingly innocent notion of simplicity in an attempt to argue against skeptical hypotheses.

sis is a function of such considerations as the number of the entities it postulates, the number of different kinds of entities it postulates, the number of laws it postulates, and the number of variables that are related in these laws.[2]

Since these various facets might be emphasized in varying degrees, this notion of simplicity is somewhat indeterminate. But it does have the advantage of capturing something of our common sense understanding of simplicity, which is what is most important for purposes here. For the point at issue is that if an account emphasizing these kinds of considerations is to be regarded as an account of epistemic reasons, it is not enough that it provide a rationale for thinking that a policy of believing simple hypotheses is likely to help us get at the truth eventually. There must be a rationale for thinking that the simpler of two hypotheses, all else being equal, is more likely to be true.

Is there such a rationale? It's notoriously difficult to come up with one. We cannot simply assume that the world is likely to be simple. Besides, even on the assumption that the world is simple, it is yet another step to assume that the best descriptions of it from our point of view would have to be simple. Nor will it do to argue that the history of the physical sciences is characterized by progressively greater predictive power and that the simplest explanation of this, to contradict half of Wilde's aphorism, is that the truth is rarely pure but usually simple. This is a tempting enough strategy. It is tempting to argue that the simplest explanation of the increasing predictive power of the physical sciences is in terms of the increasing verisimilitude (i.e., approximate truth) of physical theory. The simplest explanation of this, in turn, is that the criteria of theory choice that scientists actually use, among which (it can be argued) is simplicity, are marks of verisimilitude.

The problem with this argument is the obvious one —the explanations being invoked themselves presuppose that simplicity is a mark of verisimilitude. Thus, insofar as we are

---

[2] I take these from Richard Swinburne, "Simplicity and Choice of Theory", unpublished. Compare with Nelson Goodman, *The Structure of Appearance*, 2nd edition (Indianapolis: Bobbs-Merrill, 1966).

casting about for some rationale, even a vague one, for thinking that simplicity is somehow a mark of truth, this kind of argument isn't going to do us much good.

So, for the moment at least, I will assume that there is no rationale, or at least no non-question begging one, for thinking that simplicity is a mark of truth. Does this mean that accounts of rational belief that emphasize considerations of simplicity are inadequate? Not necessarily. It only means it is best not to interpret them as accounts of epistemic reasons —i.e., as ones that are concerned with what we have reasons to believe insofar as our aims are purely epistemic. Rather, it is best to interpret them as being concerned with what we have reasons to believe insofar we have a broader range of aims, including long-term intellectual aims and perhaps even non–intellectual aims.

Considerations of simplicity would seem to find a more natural home in this kind of account. Indeed, it seems easy to provide a plausible rationale for them —viz., simple hypotheses are easier for us to use than complex ones. They are easier to manipulate and easier to test, for example. This in turn suggests that a policy of believing the simplest adequate explanation is a policy that over the long-run will help promote our long-term intellectual goals as well as our non-intellectual goals. But if so, there are good reasons, all else being equal, to believe the simpler of two hypotheses, only these reasons are not epistemic reasons. They are reasons based on considerations of theoretical convenience, not on considerations that purport to show that the truth is apt to be simple.

This is a solution that appears to give both sides their due. The opponents of simplicity are allowed to say that we don't have an epistemic reason to believe the simplest of otherwise adequate hypotheses, while the proponents of simplicity are allowed to say that we nonetheless do have a reason to believe such hypotheses. Thus, both sides seem to get what they most want. The solution seems almost too good (and too simple) to be true. And it is.

The problem is that although simple hypotheses may be useful, this doesn't necessarily give us an adequate reason to believe them. Why not? Because we need not believe the simplest hypothesis in order to win the benefits of simplicity. Ordinarily, any intellectual or practical benefits that can be won by believing a simple hypothesis can also be won by adopting an attitude that is weaker than belief. They can be won by committing ourselves to the truth of the hypothesis. If you are a scientist, for example, you can adopt it as a working hypothesis. You can use the hypothesis without actually believing it. You can use it in the design of experiments, the formulation of other hypotheses, and so on.[3]

Admittedly, a policy of commitment may not always be effective. There may be an occasional situation in which the benefits of simple explanations, like the benefits Pascal associated with theism, can be won only with genuine belief. Perhaps only belief would generate the perseverance and optimism necessary to win the benefits. But ordinarily, a sufficiently robust attitude of commitment would also generate these benefits, and it would do so without sacrificing the accuracy of your belief system —i.e., without sacrificing your epistemic goal.

Problems of this sort arise because simplicity cannot be plausibly construed as a mark of truth, or at least so I have been assuming. This creates a dilemma for accounts of rational belief that emphasize considerations of simplicity. On the one hand, they cannot plausibly be regarded as describing what we have epistemic reasons to believe, since simplicity is not a mark of truth. But neither can they be plausibly regarded as describing what we have reasons to believe, when a broader range of our goals, both epistemic and non–epistemic, are taken into consideration. For with respect to this broader range of goals, a policy of committing ourselves to the truth of the simplest hypothesis, all else being equal, will be preferable to a policy of actually believing the hypothesis.

The most straightforward way to avoid this dilemma is to find some rationale for the idea that simplicity is a mark of truth. But for my purposes here, the important point is not so much whether there is such a rationale or not. The point

[3] Compare with Bas van Fraassen, *The Scientific Image* (Chicago: University of Chicago Press, 1985).

rather is that accounts that emphasize simplicity have no hope of fulfilling the aspirations of their proponents without some such rationale. The proponents of these accounts want considerations of simplicity to have a general applicability. There is, they think, something that recommends these considerations to us all as reasons for belief. But if this is to be so, the reasons must be epistemic ones, and for them to be epistemic, simplicity must be a mark of truth, or at least a mark of verisimilitude.

This is not a terminological remark about the term 'epistemic'. It is a point about rational belief. If considerations of simplicity are going to make it rational for us to believe something, it must be plausible to construe them as marks of truth. There must be a rationale for the idea that by believing simple theories, we increase the likelihood of our now having an accurate and comprehensive belief system. It will not do to argue that by believing such theories, we are likely to win various benefits in the future. The case must be made in terms of our present belief systems. Otherwise, the criteria will at best identify hypotheses to whose truth we have reasons to commit ourselves rather than hypotheses that we have reasons actually to believe.

At first glance, it might seem otherwise. It might seem as if considerations of simplicity could provide us with adequate reasons to believe hypotheses even if these reasons were not epistemic ones. But in fact this isn't so. If the purported benefits of believing simple hypotheses are not epistemic ones, then the best policy is one of commitment rather than belief.

There is cost to this way of thinking about simplicity, however, and it is a one that will make many philosophers uncomfortable. The cost is irrationality. For if, as I have been assuming, simplicity is not a mark of truth but we nonetheless have a tendency to believe rather than merely commit ourselves to simple hypotheses, then we are to that degree irrational. We have neither adequate evidential reasons nor adequate non-evidential reasons for these beliefs. We are thus less rational than we might have hoped.

In itself this is not an objection. What we hope for and what is the case are two different things. Still, it is enough to motivate a second look at the issue, if for no other reason than in hopes of showing that the irrationality here is not as serious as it might initially appear to be. A good place to begin this second look is with the notion of commitment.

There are various ways in which we can intellectually commit ourselves to the truth of a proposition. We can do so by presupposing it, for example, or postulating it or hypothesizing it or assuming it. Each of these is an attitude distinct from belief, but each is also distinct from merely acting as if the proposition were true. The latter can be entirely a matter of public display. Commitment cannot be. It is a deeper phenomenon, requiring a degree of intellectual engagement that need not be present when we are merely acting as if a claim were true. Commitment requires intellectual resolve, a resolve, for instance, to think about matters in a certain way.

Nevertheless, committing ourselves to the truth of a proposition and merely acting as if it were true have much in common, especially in the ways that they contrast with belief. For example, they are both context dependent whereas belief is not. When we commit ourselves to a proposition, as when we are merely acting as if it were true, we are ordinarily prepared to do so only in a limited range of situations. If you are a scientist, for example, you might commit yourself to the truth of a hypothesis within the confines of your professional life, but if you were asked about the hypothesis in a social setting, you might be reluctant to take its truth for granted. Likewise, in your deliberations about, say, a political problem to which the hypothesis is relevant, you might be reluctant to assume its truth.

Genuine belief is not like this. We don't believe a hypothesis relative to a context. We either believe it or we don't. As a result, belief is neither necessary nor sufficient for commitment. Just as you can commit yourself to a hypothesis that you do not really believe, so too you can refuse to commit yourself to a hypothesis that you do believe, since commitment might have unwelcome consequences. Perhaps it would be politically unwise. It might even be intellectually unwise. Suppose you believe in the truth of a novel hypothesis but you yourself recognize that the case to be made for it, while strong, is not yet overwhelming. Then it might be premature for you to commit yourself to it. It might be premature, for instance, to assume it in the design of experiments and the formulation of other hypotheses. Doing so might impede rather than promote its acceptance and hence impede rather than promote intellectual progress.

There are other parallels between committing ourselves to the truth of a proposition and merely acting as if it were true. Consider situations in which we have reasons to act as if some proposition were true. Our daily lives are filled with such situations. You may have reasons to act as if the tie that has just been given to you really is your favorite color, despite the fact that you avoid buying ties of that color for yourself. You may have reasons to act as if the fish that has been specially prepared for you really isn't overcooked, despite its dryness. You may have reasons to act as if your colleague's presentation really wasn't overly repetitive, despite the obvious restlessness of the audience. Fear of giving offense may give you an adequate reason to act as if you believed these propositions were true, but on the other hand it does not give you an adequate reason to believe that they are true.

Even so, it wouldn't be surprising if you had a tendency to believe such propositions and not merely a tendency to feign belief. You are touched by the gift, and perhaps this causes you to disregard, at least momentarily, the fact that you have avoided ties of this color in the past. You get caught up in the good spirits of the table, and only later does it occur to you that contrary to your sincere compliments, the fish really wasn't very good. You want your colleague to do well and this prompts you to discount the restlessness of the audience, taking it to be a consequence of the hard, uncomfortable seats rather than an indication of boredom.

With this as background, consider the role that considerations of simplicity play in shaping our beliefs. According to some, the role is pervasive. The claim is that all of us make use of considerations of simplicity in virtually all of our theorizing and that if we did not we would believe only a fraction of what we do.[4] We make use of simplicity, it is said, when we postulate atoms, electrons, and quarks but also when we try

---

[4] See Nelson Goodman, *The Structure of Appearance*, 2nd edition (Indianapolis: Bobbs-Merrill, 1966); W.V.O. Quine, "Simple Theories of a

to explain our friend's rudeness and when we refuse to take seriously the conspiracy theories of the flat earth society. A view of this sort lies behind Nelson Goodman's remark that "it is almost a commonplace that it was considerations of simplicity that led to the rejection of the Ptolemaic system".

At best this is an exaggeration. What perhaps is a commonplace is that with enough alteration, the Ptolemaic system could have been made adequate to the observations. It is also true that if the existing background theory and information had been held fixed, these alterations would have required a more complex theory. On the other hand, it is far from obvious that if clever enough changes had been made in the background theory, the resulting overall theory —i.e., the Ptolemaic theory plus the revised background theory— would necessarily have been any less simple in an objective sense. It's not obvious, in other words, that the overall theory would have required us to postulate a larger number of entities, kinds of entities, laws, variables, and so on. Of course, changes in the background theory would have required complex revisions of existing beliefs. So, in this belief-relative sense, retaining the Ptolemaic system would have been less simple, but on the other hand it also would have been lacking in credibility relative to these beliefs.

For similar reasons, it is at best an exaggeration to claim that considerations of simplicity have a pervasive role in everyday theorizing. If you have limited the number of hypotheses you are seriously considering to just two and if neither is terribly complex, then considerations of simplicity probably will not play much of a role in your deliberations about the respective merits of these two hypotheses, even if one is somewhat more simple than the other.

But even if some philosophers have exaggerated its importance in our reasoning, it is clear that simplicity sometimes does play a role in shaping our beliefs, although the role may be a less fine-grained and more negative one than is often

---

Complex World"; Richard Swinburne, "Simplicity and Choice of Theory"; William Lycan, *Judgement and Justification* (Cambridge: Cambridge University Press, 1988); Elliot Sober, *Simplicity* (Oxford: Clarendon Press, 1975).

thought. At the very least, considerations of simplicity play an indirect role role in theorizing. They do so if for no other reason than because we have only limited cognitive abilities and only a limited amount of time to exercise these abilities. Indeed, if a hypothesis is complex enough, it may not be possible for us even to understand it. So, it won't even be a candidate for belief. It will be filtered out automatically. But even among those hypotheses that we are able to understand, some will be so complex that it would be impractical for us to take them seriously. It would take far too much of our time even to deliberate about them, much less to use them in making predictions and constructing other hypotheses. They are so complex that they would be of little value to us even if they were true. Thus, they too will be filtered out. We won't take them seriously. We will simply ignore them.

This need not be a conscious policy on our part. It's not as if we have made a decision to follow such a procedure. It's more likely to be a matter of our dispositions and practices, ones that most of us have not given much thought. But the result of these dispositions and practices is that simple hypotheses as a group tend to get preferential treatment over complex hypotheses as a group. As a group, we are disposed to take them more seriously than complex ones, and often enough the result may be that we are more disposed to believe them as well.

We are disposed to believe them even though we may not have adequate reason to believe them. The simplicity of a hypothesis, I am assuming, does not give us a reason to believe it insofar as our aims are purely epistemic. Moreover, we need not have any other adequate epistemic reasons to believe them either. Whether or not we do depends in part on how extensive a role simplicity has played in picking out the hypothesis. The more extensive the role, the less likely it is that we have adequate epistemic reasons to believe it. Nor need we have good reasons to believe it when a broader range of our ends, both epistemic and non-epistemic, are taken into consideration. At most we may have reasons only to commit ourselves to its truth.

Even so, none of this may not be worth worrying about. It may be no more worth worrying about than your belief about the tie. Believing that the tie really is your favorite color encourages you to act in a way that does not give offense, and even though you lack adequate evidence for this proposition, your believing it need not have widespread detrimental effects. Of course, we can imagine situations in which it would have unwelcome effects; it might somehow prompt you to make a bad decision about some other, more important matter. But this need not be the case, and presumably it often is not.

Something similar may be true of our disposition to believe simple hypotheses. Even if we do not have adequate evidence for these hypotheses and even if the benefits that accrue to us from these beliefs could have been secured by committing ourselves to the hypotheses instead, none of this may matter very much. As long as these beliefs do not adversely affect the rest of our lives, as they ordinarily will not, there is no need for great concern.

This will not be enough to satisfy those who think that believing the simplest of otherwise equal hypotheses is positively desirable. For them, it is not enough to point out that having such beliefs is not harmful. There must also be a positive rationale, a rationale for thinking that believing simple hypotheses is preferable to a policy of committing ourselves to them.

The problem is that there doesn't seem to be any such rationale. Or at least this is so if we continue to assume that simplicity is not a mark of truth. Given this assumption, there is no way to provide the rationale in terms of a purely epistemic goal, and there seems to be no way of doing so in terms of our practical or long-term intellectual goals either. These goals can be won, and at no epistemic cost, by committing ourselves to the truth of simple hypotheses rather than genuinely believing them.

Even so, the point I want to emphasize is that the irrationality involved in believing simple hypotheses, if that be what it is, may be of a very weak kind. One way to appreciate this is to distinguish the beliefs it is irrational for us to acquire from the beliefs it is irrational for us to retain. Even if you did not have adequate reason to acquire a belief, it need not be rational for you to go to the trouble of ridding yourself of it, given that you do have it. This may be the case for belief in a simple hypothesis. If no significant harm is done by your believing rather than merely committing yourself to the hypothesis, it may not be irrational for you to continue believing it. This need not be irrational even though you don't have adequate evidence for it and even though you had no other adequate reason to begin believing it either. Going to the trouble of ridding yourself of the belief may not be worth the effort.

Moreover, you may not even have an adequate reason to try to change yourself so that you will not believe simple hypotheses in the future. After all, doing so may not be easy. If you have an inclination to believe simple hypotheses, then given that simplicity is not a mark of truth, you have a disposition to be irrational —i.e., a disposition to believe hypotheses that you do not have adequate reasons to believe. Nevertheless, it might not be rational for you to try to rid yourself of this disposition. Doing so might be too costly. It may not be worth trouble, especially if the disposition is a deep-seated one.

Neither of these points imply that it is rational for you to believe as opposed to commit yourself to the simplest of otherwise equal hypotheses. On the contrary, both acknowledge that it is not rational for you to do so. However, they do take much of the sting out of this kind of irrationality. They suggest that it can be rational for you to put up with this kind of irrationality in yourself.

An even stronger and more interesting result may also be possible. The set of cognitive dispositions that is best for you, given the kind of creature that you are, may include a disposition to believe rather than merely commit yourself to simple hypotheses. All other sets, or at least all those that are genuine possibilities for you, might be worse than this set. They might be worse even though this set inclines you to acquire one kind of belief that you lack adequate reasons for acquiring —belief in simple hypotheses. Even so, all the other sets might have more serious drawbacks, ones that would make your overall situation worse. If so, it may be rational for you to have the set of dispositions that you now have, despite the fact that these dispositions tend to produce some irrational beliefs. In other words, it may be rational for you to be irrational in this way.

This has the sound of paradox, but it's merely a recognition that in your intellectual life, as in the rest of your life, you are operating under various kinds of constraints. Consider an analogy. The love you have for your family may sometimes cause you to act wrongly. It may cause you to favor excessively your family's interests over the interests of others. You go too far in your efforts to do what is best for your family, and the result is that sometimes your actions cause undue hardships for others. Still, it might be morally worse if you did not have this strong love for your family. If you loved them less, you perhaps wouldn't be disposed to favor them excessively. So, it would correct this problem, but on the other hand it might have bad consequences for your family. If you loved them any less, you might carry out your responsibilities towards them less well, and this might lead to results that overall are morally much worse than those that are produced by your being too zealous in favoring them. Thus, it might be best for you to have this strong love for your family. This might be best even though, given the kind of person you are, it sometimes inclines you to act wrongly.[5]

Similarly, given the kind of cognitive creature you are, it may be positively desirable for you to have a disposition to believe simple hypotheses. It might be part of the best set of cognitive dispositions for you to have. Any other set, or at least any other set that you are capable of, might be worse. If you didn't have this disposition, you wouldn't be as inclined to believe simple hypotheses and, thus, you would avoid one kind of irrational belief. But as in the case where your love for your family inclines you to act wrongly, it might have other drawbacks. Were you to lack this disposition, perhaps you would find yourself spending more time in deliberations over whether to believe or merely commit yourself to hypotheses, and this might divert your attention from other, more important matters, causing you to make errors of judgement about them. If so, it is rational for you to have this disposi-

[5] See Derek Parfit, *Reasons and Persons* (Oxford: Oxford University Press, 1986), especially Chapter 1.

tion, despite the fact that it inclines you towards irrational beliefs.[6]

Points of this general sort are enormously important for epistemology. They mark the path towards a realistic epistemology, one that pays something more than lip service to the idea that our intellectual lives are constrained by the kind of cognitive creatures we are.

In doing so, points of this sort also provide a useful framework for thinking about the significance of recent empirical work that has been done on irrational belief. Much of this work seems to show that we have disturbingly stubborn tendencies to make certain kinds of faulty inferences —e.g., even well educated subjects tend not to seek out relevant and readily available base rate information.[7] But points of the above sort hold out at least some hope that the cognitive dispositions that produce these faulty inferences might nonetheless be rational ones for us to have. They might be rational because they might serve us well enough on most occasions. Indeed, they might even be part of the best set of cognitive dispositions available to us. Any other available set might have even worse failings. Or short of this, they at least may be dispositions that it is not rational for us to go to the trouble of extinguishing. In our normal everyday affairs, they may not do enough harm to make it rational for us to try to rid ourselves of them. Rather, it is rational for us to put up with them.

But this is a side issue. For my purposes here, the importance of such points is that they constitute the beginnings of a possible defense for simplicity. Moreover, similar defenses may be available for many of the other theoretical virtues —fertility and problem-solving effectiveness, for example. It is not a defense that tries to argue that simplicity really is a

---

[6] Similarly, if your preference for simple hypotheses is a matter of deliberate policy, this policy might be a rational one for you to follow even though beliefs in accordance with it are often irrational. This is so because it might nonetheless be part of your best set of policies. Any other set of policies, or at least any other that you are capable of following, might leave you worse off.

[7] For a quick overview of some of this work, see Stephen Stich, The Fragmentation of Reason (Cambridge: MIT Press, 1990).

mark of truth or that all of us at least implicitly treat it as such. Nor is it a defense that ignores the difference between belief and commitment. On the contrary, it acknowledges that it may very well be irrational for us to believe as opposed to commit ourselves to a hypothesis if simplicity has played a significant role in selecting it. This may be irrational not only in an epistemic sense, but in a broader sense as well —i.e., a sense that takes into consideration our long-term intellectual goals and our practical goals.

The defense instead consists of arguing that the irrationality here is of a very weak kind. A mark of its weakness is that ordinarily we won't have adequate reasons to rid ourselves of the belief in the hypothesis, and likewise we ordinarily won't have adequate reasons to rid ourselves of the disposition to acquire such beliefs. On the contrary, it may even be positively rational for us to have such a disposition, despite the fact that it often inclines us to have irrational beliefs. It may be rational for us to be irrational in this way.

12

PHILOSOPHICAL ISSUES, 3
Science and Knowledge, 1993

Folk Psychology: Simulation or Tacit Theory?

Stephen Stich and Shaun Nichols

1 Introduction

A central goal of contemporary cognitive science is the explanation of cognitive abilities or capacities. [Cummins 1983] During the last three decades a wide range of cognitive capacities have been subjected to careful empirical scrutiny. The adult's ability to produce and comprehend natural language sentences and the child's capacity to acquire a natural language were among the first to be explored. [Chomsky 1965, Fodor, Bever & Garrett 1974, Pinker 1989] There is also a rich literature on the ability to solve mathematical problems [Greeno 1983], the ability to recognize objects visually [Rock 1983, Gregory 1970, Marr 1982], the ability to manipulate and predict the behavior of middle sized physical objects [McClosky 1983, Hayes 1985], and a host of others.

In all of this work, the dominant explanatory strategy proceeds by positing an internally represented "knowledge structure" —typically a body of rules or principles or propositions— which serves to guide the execution of the capacity to be explained. These rules or principles or propositions are often described as the agent's "theory" of the domain in question. In some cases, the theory may be partly accessible to consciousness; the agent can tell us some of the rules or principles he is using. More often, however, the agent has no conscious access to the knowledge guiding his behavior. The theory is "tacit" [Chomsky 1965] or "sub–doxastic" [Stich 1978]. Perhaps the earliest philosophical account of this explanatory strategy is set out in Jerry Fodor's paper, "The Appeal to Tacit Knowledge in Psychological Explanations". [Fodor 1968]. Since then the idea has been elaborated by Dennett [1978a], Lycan [1981, 1988], and a host of others.

Among the many cognitive capacities that people manifest, there is one cluster that holds a particular fascination for philosophers. Included in this cluster is the ability to *describe* people and their behavior (including their linguistic behavior) *in intentional terms* —or to "interpret" them, as philosophers sometimes say. We exercise this ability when we describe John as *believing that the mail has come*, or when we say that Anna *wants to go to the library.* By exploiting these intentional descriptions, people are able to offer *explanations* of each other's behavior (Susan left the building *because* she believed that it was on fire) and to *predict* each other's behavior, often with impressive accuracy. Since the dominant strategy for explaining any cognitive capacity is to posit an internally represented theory, it is not surprising that in this area, too, it is generally assumed that a theory is being invoked. [Churchland 1981 & 1989, Fodor 1987, Sellars 1963. See also Olson et. al. 1988] The term "folk psychology" has been widely used as a label for the largely tacit psychological theory that underlies these abilities. During the last decade or so there has been a fair amount of empirical work aimed at describing or modeling folk psychology and tracking its emergence and development in the child. [D'Andrade 1987, Leslie 1987, Astington et al. 1988]

Recently, however, Robert Gordon, Alvin Goldman and a number of other philosophers have offered a bold challenge to the received view about the cognitive mechanisms underlying our ability to describe, predict and explain people's behavior. [Goldman 1989, Gordon 1986, Gordon unpublished a, Gordon unpublished b, Montgomery 1987, Ripstein 1987, Heal 1986][1] Though they differ on the details, these philosophers agree in denying that an internally represented folk psychological theory plays a central role in the exercise of these abilities. They also agree that a special sort of mental *simulation* in which we use ourselves as a model for the person we are describing or predicting, will play an important role in the correct account of the mechanisms subserving these abilities. In this paper, although we will occasionally mention the views of other advocates of simulation, our principal focus will be on Gordon and Goldman.

If these philosophers are right, two enormously important consequences will follow. First, of course, the dominant explanatory strategy in cognitive science, the strategy that appeals to internally represented knowledge structures, will be shown to be mistaken in at least one crucial corner of our mental lives. And if it is mistaken there, then perhaps theorists exploring other cognitive capacities can no longer simply take the strategy for granted.

To explain the second consequence we will need a quick review of one of the central debates in recent philosophy of mind. The issue in the debate is the very existence of the intentional mental states that are appealed to in our ordinary explanations of behavior —states like believing, desiring, thinking, hoping, and the rest. *Eliminativists* maintain that there really are no such things. Beliefs and desires are like phlogiston, caloric and witches; they are the mistaken posits of a radically false theory. The theory in question is "folk psychology" —the collection of psychological principles and generalizations which, according to eliminativists (and most of their opponents) underlies our everyday explanations of behavior. The central premise in the eliminativist's argument is that neuroscience (or connectionism or cognitive science) is on the verge of demonstrating persuasively that

---

[1] We are grateful to Professor Gordon for providing us with copies of his unpublished papers, and for allowing us to quote from them at some length.

folk psychology is false. But if Gordon and Goldman are right, they will have pulled the rug out from under the eliminativists. For if what underlies our ordinary explanatory practice is not a theory at all, then obviously it cannot be a radically false theory. There is a certain delightful irony in the Gordon/Goldman attack on eliminativism. Indeed, one might almost view it as attempting a bit of philosophical jujitsu. The eliminativists claim that there are no such things as beliefs and desires because the folk psychology that posits them is a radically false theory. Gordon and Goldman claim that the theory which posits a tacitly known folk psychology is *itself* radically false, since there are much better ways of explaining people's abilities to interpret and predict behavior. Thus, if Gordon and Goldman are right, *there is no such thing as folk psychology!* [Gordon 1986, p. 170; Goldman 1989, p. 182]

There can be no doubt that if Gordon and Goldman are right, then the impact on both cognitive science and the philosophy of mind will be considerable. But it is a lot easier to doubt that their views about mental simulation are defensible. The remainder of this paper will be devoted to developing these doubts. Here's the game plan for the pages to follow. In Sections 2 and 3, we will try to get as clear as we can on what the simulation theorists claim. We'll begin, in Section 2, with an account of the special sort of simulation that lies at the heart of the Gordon/Goldman proposal. In that section our focus will be on the way that simulation might be used in the *prediction* of behavior. In Section 3, we'll explore the ways in which mental simulation might be used to explain the other two cognitive capacities that have been of special interest to philosophers: *explaining* behavior and producing *intentional descriptions* or *interpretations*. We'll also consider the possibility that simulation might be used in explaining the *meaning* of intentional terms like 'believes,' and 'desires'. Since the accounts of simulation that Gordon and Goldman have offered have been a bit sketchy, there will be a lot of filling in to do in Sections 2 and 3. But throughout both Sections, our goal will be sympathetic interpretation; we've tried hard not to build straw men. In the following two Sections, our stance turns critical. In Section 4, we will do our best to assemble all the arguments offered by Gordon and Goldman in support of their simulation theory, and to explain why none of them are convincing. In Section 5 we will offer two arguments of our own, aimed at showing why, in light of currently available evidence, the simulation theory is very implausible indeed. Section 6 is a brief conclusion.

2 Predicting Behavior: Theory, Simulation and Imagination

Suppose that you are an aeronautical engineer and that you want to predict how a newly built plane will behave at a certain speed. There are two rather different ways in which you might proceed. One way is to sit down with pencil and paper, a detailed set of specifications of the plane, and a state of the art textbook on aerodynamic theory, and try to calculate what the theory entails about the behavior of the plane. Alternatively, you could build a model of the plane, put it in a wind tunnel, and observe how it behaves. You have to use a bit of theory in this second strategy, of course, since you have to have some idea which properties of the plane you want to duplicate in your model. But there is a clear sense in which a theory is playing the central role in the first prediction and a model or simulation is playing a central role in the second.[2]

Much the same story could be told if what you want to do is predict the behavior of a person. Suppose, for example, you want to predict what a certain rising young political figure would do if someone in authority tells him to administer painful electric shocks to a person strapped in a chair in the next room. One approach is to gather as much data as you can about the history and personality of the politician and then consult the best theory available on the determinants of behavior under such circumstances. Another approach is

[2]The wind tunnel analogy is suggested by Ripstein (1987 p. 475 ff.). Gordon also mentions the analogy (unpublished, b, p. 8) but he puts it to a rather different use.

to set up a Milgram–style experiment and observe how some other people behave. Naturally, it would be a good idea to find experimental subjects who are psychologically similar to the political figure whose behavior you are trying to predict. Here, as before, theory plays a central role in the first prediction, while a simulation plays a central role in the second.

In both the aeronautical case and the psychological case, we have been supposing that much of the predicting process is carried on outside the predictor. You do your calculations on a piece of paper; your simulations are done in wind tunnels or laboratories. But, of course, it will often be possible to internalize this process. The case is clearest when a theory is being used. Rather than looking in a textbook, you could memorize the theory, and rather than doing the calculations on a piece of paper, you could do them in your head. Moreover, it seems entirely possible that you could learn the theory so well that you are hardly conscious of using it or of doing any explicit calculation or reasoning. Indeed, this, near enough, is the standard story about a wide variety of cognitive capacities.

A parallel story might be told for predictions using simulations. Rather than building a model and putting it in a wind tunnel, you could *imagine* the model in the wind tunnel and see how your imaginary model behaves. Similarly, you could *imagine* putting someone in a Milgram–style laboratory and see how your imaginary subject behaves. But obviously there is a problem lurking here. For while it is certainly possible to imagine a plane in a wind tunnel, it is not at all clear how you could successfully imagine the behavior of the plane unless you had a fair amount of detailed information about the behavior of planes in situations like this one. When the simulation uses a real model plane, the world tells you how the model will behave. You just have to look and see. But when you are only imagining the simulation, there is no real model for you to look at. So it seems that you must have an internalized knowledge structure to guide your imagination. The theory or knowledge structure that you are exploiting may, of course, be a tacit one, and you may be quite unaware that you are using it. But unless we suppose your imagination is guided by some systematic body of information about the behavior of planes in situations like this one, the success of your prediction would be magic.

When you are imagining the behavior of a person, however, there are various ways in which the underlying system might work. One possibility is that imagining the behavior of a person is entirely parallel to imagining the behavior of a plane. In both cases your imagination is guided by a largely tacit theory or knowledge structure. But there is also a very different mechanism that might be used. In the plane case, you don't have a real plane to observe, so you have to rely on some stored information about planes. You do, however, have a real, human cognitive system to observe —your own. Here's a plausible, though obviously over–simplified, story about how that system normally works:

At any given time you have a large store of beliefs and desires. Some of the beliefs are derived from perception, others from inference. Some of the desires (like the desire to get a drink) arise from systems monitoring bodily states, others (like the desire to go into the kitchen) are "sub–goals" generated by the decision making (or "practical reasoning") system. The decision making system, which takes your beliefs and desires as input, does more than generate sub–goals, it also somehow or other comes up with a decision about what to do. That decision is then passed on to the "action controllers" —the mental mechanisms responsible for sequencing and coordinating the behavior necessary to carry out the decision. Rendered boxologically, the account just sketched appears in Figure 1.

Now suppose that it is possible to take the decision making system "off–line" by disengaging the connection between the system and the action controllers. You might then use it to generate decisions that you are not about to act on. Suppose further that in this off–line mode, you can feed the decision making system some hypothetical or "pretend" beliefs and desires —beliefs and desires that you do not actually have, but that the person whose behavior you're trying to predict does. If all this were possible, you could then sit back and let the system generate a decision. Moreover, if your decision making system is similar to the one in the person whose

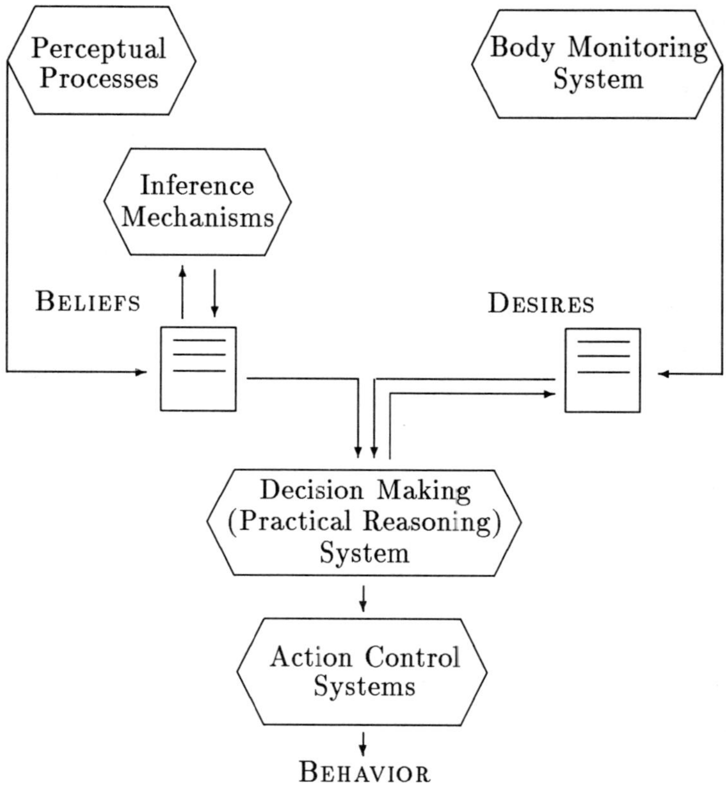

FIGURE 1.

behavior you're trying to predict, and if the hypothetical beliefs and desires you've fed into your system off–line are close to the ones that he has, then the decision that your system generates will often be similar to the one that his system generates. There is no need for a special internalized knowledge structure here; no tacit folk psychological theory is being used. Rather, you are using (part of) your own cognitive mechanism as a model for (part of) his. Moreover, just as in the case where the prediction exploits a theory, this whole process may be largely unconscious. It may be that all you are aware of is the prediction itself. Alternatively, if you consciously imagine what the target of your prediction will do, it could well be the case that your imagination is guided by this simulation rather than by some internally represented psychological theory.

We now have at least the outline of an account of how mental simulation might be used in predicting another person's behavior. An entirely parallel story can be told about predicting our own behavior under counterfactual circumstances. If, for example, I want to know what I would I do if I believed that there was a burglar in the basement, I can simply take my decision making system off–line and provide it with the pretend belief that there is a burglar in the basement.[3]

In the next section we'll try to get clear on how this process of simulation might be used in explaining various other cognitive capacities. But before attending to that task, we would do well to assemble a few quotes to confirm our claim that the story we've told is very close to the one that those we'll be criticizing have in mind. Gordon is much more explicit than Goldman on the use of simulation in prediction. Here's a passage from his 1986 paper:

> [O]ur decision–making or practical reasoning system gets partially disengaged from its 'natural' inputs and fed instead with suppositions and images (or their 'subpersonal' or 'sub–doxastic' counterparts). Given these artificial pretend inputs the system then 'makes up its mind' what to do. Since the system is being run off–line, as it were, disengaged also from its natural output systems, its 'decision' isn't actually executed but rather ends up as an anticipation ... of the other's behavior. [Gordon 1986, p. 170]

And another, this time from an unpublished manuscript contrasting his view to Fodor's:

> The Simulation Theory as I present it holds that we explain and predict behavior not by applying a theory but simply by exercising a skill that has two components: the capacity for practical reasoning —roughly, for making decisions on the basis of facts

[3] The burglar in the basement example is borrowed from Gordon 1986, p. 161.

> and values— and the capacity to introduce "pretend" facts and values into one's decision–making typically to adjust for relevant differences in situation and past behavior. One predicts what the other will decide to do by making a decision oneself —a "pretend" decision, of course, made only in imagination— after making such adjustments. [Gordon unpublished, a, MS p. 3]

Gordon later suggests that the capacity to simulate in this way may be largely innate:

> [Evidence] suggests that the readiness for simulation is a prepackaged "module" called upon automatically in the perception of other human beings.[4] It suggests also that supporting and complementing the conscious, reportable procedure we call putting ourselves in the other's place, those neural systems that are responsible for the formation of emotions and intentions are, often without our knowledge, allowed to run off–line: They are partially disengaged from their "natural" inputs from perception and memory and fed artificial pretend inputs; uncoupled also from their natural output systems, they terminate not as intentions and emotions but as anticipations of, or perhaps just unconscious motor adjustments to, the other's intentions, emotions, behavior. [Gordon unpublished a, MS p. 5]

3 Other Uses For Simulation: Explanation, Interpretation and the Meaning of Intentional Terms

Let's turn, now, to people's ability to offer *intentional explanations* of other people's actions. How might mental simulation be used to account for that ability?

Consider, for example, a case similar to one proposed by Gordon.[5] We are seated at a restaurant and someone comes

[4]The evidence Gordon cites includes the tendency to mimic other people's facial expressions and overt bodily movements, and the tendency in both humans and other animals to direct one's eyes to the target of a conspecific's gaze.

[5]Gordon 1986, p. 163 ff.

up to us and starts speaking to us in a foreign language. How might simulation be exploited in producing an intentional explanation for that behavior?

One proposal, endorsed by both Gordon and Goldman, begins with the fact that simulations can be used in predictions, and goes on to suggest that intentional explanations can be generated by invoking something akin to the strategy of analysis–by–synthesis. In using simulations to predict behavior, hypothetical beliefs and desires are fed into our own decision making system (being used "off–line" of course), and we predict that the agent would do what we would decide to do, given those beliefs and desires. A first step in *explaining* a behavioral episode that has already occurred is to see if we can find some hypothetical beliefs and desires which, when fed into our decision mechanism, will produce a decision to perform the behavior we want to explain.

Generally, of course, there will be *lots* of hypothetical beliefs and desires that might lead us to the behavior in question. Here are just a few:

(a) If we believe someone only speaks a certain foreign language and we want to ask him something, then we would decide to speak to him in that language.

(b) If we want to impress someone and we believe that speaking in a foreign language will impress him, then we will decide to speak to him in that language.

(c) If we believe that speaking to someone in a foreign language will make him laugh, and if we want to make him laugh, then we will decide to speak to him in that language.

And so on. Each of these simulation–based predictions provides the kernel for a possible explanation of the behavior we are trying to explain. To decide among these alternative explanations, we must determine which of the input belief/desire pairs is most plausibly attributed to the agent. Some belief/desire pairs will be easy to exclude. Perhaps the agent is a dour fellow; he never wants to make anyone laugh. If we believe this to be the case, then (c) won't be very plausible. In other cases we can use information about the agent's perceptual situation to assess the likelihood of various beliefs. If Mary has just made a rude gesture directly in front of the agent, then it is likely the agent will believe that Mary has insulted him. If the rude gesture was made behind the agent's back, then it is not likely he will believe that she has insulted him. In still other cases, we may have some pre–existing knowledge of the agent's beliefs and desires. But, as both Goldman and Gordon note, it will often be the case that there are lots of alternative explanations that can't be excluded on the basis of evidence about the agent's circumstances or his history. In these cases, Goldman maintains, we simply assume that the agent is psychologically similar to us —we attribute beliefs that are "natural for us" [Goldman 1989, p. 178] and reject (or perhaps do not even consider) hypotheses attributing beliefs that we consider to be less natural. [Goldman 1989, pp. 178-9] Gordon tells much the same story.

> No matter how long I go on testing hypotheses, I will not have tried out all candidate explanations of the [agent's] behavior. Perhaps some of the unexamined candidates would have done at least as well as the one I settle for, if I settle: perhaps indefinitely many of them would have. But these would be 'far fetched', I say intuitively. Therein I exhibit my inertial bias. The less 'fetching' (or 'stretching', as actors say) I have to do to track the other's behavior, the better. I tend to *feign* only when necessary, only when something in the other's behavior doesn't fit. This inertial bias may be thought of as a 'least effort' principle: the 'principle of least pretending'. It explains why, other things being equal, I will prefer the less radical departure from the 'real' world —i.e. from what I myself take to be the world. [Gordon 1986, p. 164][6]

---

[6] Ripstein's account of role of simulation in intentional explanation is quite similar:

> I wish to defend the claim that imagining what it would be like to be in "someone else's shoes" can serve to explain that person's actions.... I shall argue that imagining oneself in someone else's situation... allows actions to be explained without recourse to a theory of human behavior. [Ripstein 1987, p. 465]
>
> [T]he same sort of modeling [that engineers use when they study bridges in wind tunnels] is important to commonsense psychology. I

Though the views endorsed by Gordon and Goldman are generally very similar, the two writers do differ in their emphasis. For Gordon, prediction and explanation loom large, while for Goldman, the capacity to *interpret* people, or to describe them in intentional terms, is given pride of place. Part of the story Goldman tells about simulation–based intentional description relies on the account of simulation–based explanation that we have just sketched. One of the ways we determine which beliefs and desires to attribute to people is by observing their behavior and then attributing the intentional states that best explain their behavior. A second simulation– based strategy for determining which beliefs and desires to attribute focuses on the agent's perceptual situation and on his or her "basic likings or cravings". [Goldman 1989, p. 170]

> From your perceptual situation, I infer that you have certain perceptual experiences or beliefs, the same ones I would have in your situation. I may also assume (pending information to the contrary) that you have the same basic likings that I have: for food, love, warmth, and so on. [Goldman 1989, p. 170]

As we read them, there is only one important point on which Gordon and Goldman actually *dis*agree. The accounts of simulation–based prediction, explanation and interpretation that we have sketched all seem to require that the person doing the simulating must already understand intentional notions like belief and desire. A person can't pretend he believes that the cookies are in the cookie jar unless he understands what it is to believe that the cookies are in the cookie jar; nor

---

can use my personality to model yours by "trying on" various combinations of beliefs, desires and character traits. In following an explanation of what you do, I use my personality to determine that the factors mentioned would produce the result in question.... I do not need to know how you work because I can rely on the fact that I work in a similar way. My model... underwrites the explanation by demonstrating that particular beliefs and character traits would lead to particular actions under normal circumstances. [Ripstein 1987, pp. 476-477]

can a person imagine that she wants to make her friend laugh unless she understands what it is to want to make someone laugh. Moreover, as Goldman notes, when simulation is used to attribute intentional states to agents, it "assumes a prior understanding of what state it is that the interpreter attributes to [the agent]". [Goldman 1989, p. 182] Can the process of simulation somehow be used to explain the meaning or truth conditions of locutions like "*S* believes that *p*" and "*S* desires that *q*"? Goldman is skeptical, and tells us that "the simulation theory looks distinctly unpromising on this score". [Goldman 1989, p. 182] But Gordon is much more sanguine. Building on earlier suggestions by Quine, Davidson and Stich, he proposes the following account:

> My suggestion is that
>
> (2)[*Smith believes that* Dewey won the election.]
>
> be read as saying the same thing as
>
> (1)[Let's do a Smith simulation. Ready? *Dewey won the election.*]
>
> though less explicitly. [Gordon 1986, p. 167]

We are not at all sure we understand this proposal, and Gordon himself concedes that "the exposition and defense of this account of belief are much in need of further development". [Gordon 1986, p. 167] But no matter. We think we do understand the simulation–based accounts of prediction, explanation and interpretation that Gordon and Goldman both endorse. We're also pretty certain that none of these accounts is correct. In the sections that follow, we will try to say why.

4 Arguments in Support of Simulation–Based Accounts

In this Section we propose to assemble all the arguments we've been able to find in favor of simulation–based accounts and say why we don't think any of them is persuasive. Then, in the following Section, we will go on to offer some arguments of our own aimed at showing that there is lots of evidence that simulation–based accounts cannot easily accommodate, though more traditional theory–based accounts can. Before turning to the arguments, however, we would do well to get a bit clearer about the questions that the arguments are (and are not) intended to answer.

The central idea in the accounts offered by Gordon and Goldman is that in predicting, explaining or interpreting other people we simulate them by using part of *our own* cognitive systems "off–line". There might, of course, be other kinds of simulation in which we do not exploit our own decision making system in order to model the person we are simulating. But these other sorts of simulation are not our current concern. To avoid confusion, we will henceforth use the term *off–line simulation* for the sort of simulation that Gordon and Goldman propose. The question in dispute, then, is whether off–line simulation plays a central role in predicting, explaining or interpreting other people. Gordon and Goldman say yes; we say no.

It would appear that the only serious alternatives to the off–line simulation story are various versions of the "theory–theory" which maintain that prediction, explanation and interpretation exploit an internally represented theory or knowledge structure —a tacitly known "folk psychology". So if an advocate of off–line simulation can mount convincing arguments against the theory–theory, then he can reasonably claim to have made his case. The theory–theory is not the only game in town, but it is the only *other* game in town. It is not surprising, then, that in defending off–line simulation Gordon and Goldman spend a fair amount of time raising objections to the theory–theory.

There are, however, some important distinctions to be drawn among different types of theory–theories. Until fairly recently, most models that aimed at explaining cognitive capacities posited internally represented knowledge structures that invoked explicit rules or explicit sentence–like principles. But during the last decade there has been a growing dissatisfaction with sentence–based and rule–based knowledge structures, and a variety of alternatives have been explored. Perhaps the most widely discussed alternatives are

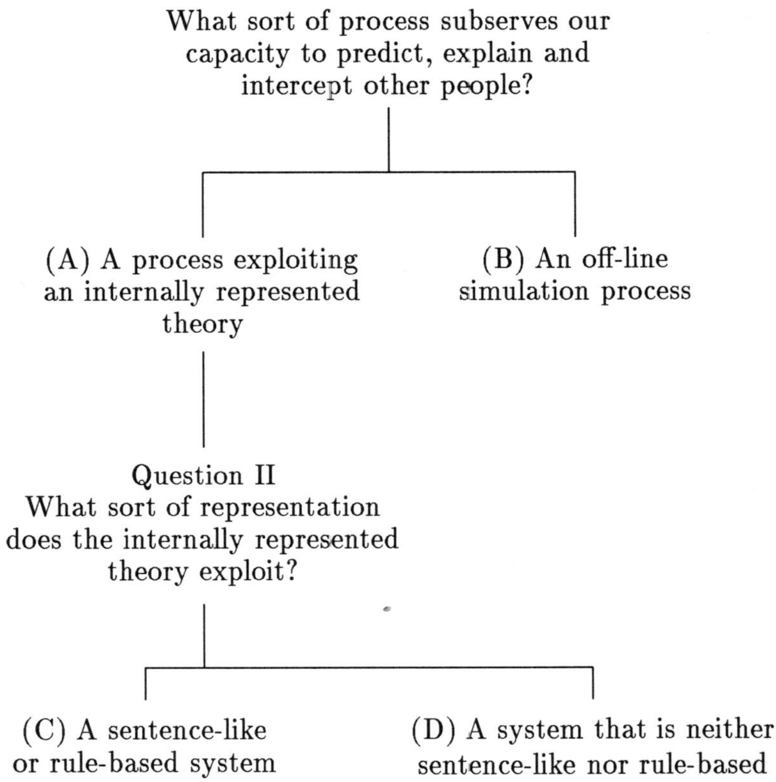

FIGURE 2.

connectionist models in which the knowledge used in making predictions is stored in the connection strengths between the nodes of a network. In many of these systems it is difficult or impossible to view the network as encoding a set of sentences or rules. [Ramsey, Stich & Garon 1990] Other theorists have proposed quite different ways in which non–sentential and non–rule–like strategies could be used to encode information. [See, for example, Johnson–Laird 1983]

Unfortunately, there is no terminological consensus in this domain. Some writers prefer to reserve the term "theory" for sentence–like or rule–based systems. For these writers, most connectionist models do not invoke what they would call an internally represented theory. Other writers are more liberal in their use of "theory", and are prepared to count just about any internally stored body of information about a given domain as an internally represented theory of that domain. For these writers, connectionist models and other non–sentential models do encode a tacit theory. We don't think there is any substantive issue at stake here. But the terminological disagreements can generate a certain amount of confusion. Thus, for example, someone who used "theory" in the more restrictive way might well conclude that if a connectionist (or some other non–sentence based) account of our ability to predict other people's behavior turns out to be the right one, then the theory–theory is mistaken. So far, so good. But it is important to see that the falsity of the theory–theory (narrowly construed) is no comfort at all to the off–line simulation theorist. The choice between off–line simulation theories and theory–theories is plausibly viewed as exhaustive only when "theory" is used in the wide rather than the restrictive way. For the remainder of this paper, we propose to adopt the wide interpretation of "theory". Using this terminology, the geography of the options confronting us are represented in Figure 2.[7] In the pages that follow, we will be defending option (A) in answer to Question (I). We take

[7] As Jerry Fodor has pointed out to us, the logical geography is actually a bit more complex than Figure 2 suggests. To see the point, consider the box labeled "Decision Making (Practical Reasoning) System" in Figure 1. Gordon and Goldman tell us relatively little about the contents of this box. They provide no account of how the Practical Reasoning System goes about the job of producing decisions from beliefs and desires. However, there are some theorists —Fodor assures us that he is one— who believe that the Practical Reasoning System goes about its business by exploiting an internally represented decision theory. If this is right, then we exploit a tacit theory each time we make a decision based on our beliefs and desires. But now if we make predictions about other people's behavior by taking our own Practical Reasoning System off–line, then we also exploit a tacit theory when we make these predictions. Thus, contrary to the suggestion in Figure 2, off–line simulation processes and processes exploiting an internally represented theory are not mutually exclusive, since some off–line simulation processes may also exploit a tacit theory.

In the pages that follow, we propose to be as accommodating as possible to our opponents and to make things as hard as possible for ourselves.

no stand at all on question (II). So much for getting clear on the questions. Now let's turn to the arguments.

Argument 1: No one has been able to state the principles of the internally represented folk psychological theory posited by the theory–theory.

Both Goldman and Gordon go on at some length about the fact that it has proven very difficult to state the principles or laws of the folk psychological theory that, according to the theory–theorist, guide our interpretations and predictions.

> [A]ttempts by philosophers to articulate the putative laws or 'platitudes' that comprise our folk theory have been notably

---

It is our contention that prediction, explanation and interpretation of the sorts we have discussed do not use an off–line simulation process, period. So if it turns out that Fodor is right (because the Practical Reasoning System embodies an internally represented theory) and that Gordon and Goldman are right (because we predict and explain by taking this system off–line), then we lose, and they win. Also, of course, if Fodor is wrong about how the Practical Reasoning System works but Gordon and Goldman are right about prediction, explanation and interpretation, again we lose and they win. So as we construe the controversy, it pits those who advocate any version of the off–line simulation account against those who think that prediction, explanation and interpretation are subserved by a tacit theory stored somewhere other than in the Practical Reasoning System. But do keep in mind that we interpret "theory" broadly. So, for example, if it turns out that there is some non–sentence–like, non–rule–based module which stores the information that is essential to folk psychological prediction and explanation, and if this module is not used at all in ordinary "on–line" practical reasoning and decision making, then we win and they lose.

It might be protested that in drawing the battle lines as we propose to draw them, we are conceding to the opposition a position that they never intended to occupy. As we have already noted, Gordon and Goldman expend a fair amount of effort arguing that a tacit theory is not exploited in folk psychological prediction and explanation. Since they think that the Practical Reasoning System is exploited in folk psychological prediction and explanation, presumably they would deny that the Practical Reasoning System uses an internally represented decision theory. So it is a bit odd to say that they win if Fodor is right about the Practical Reasoning System and they are right about off–line simulation. This is a point we happily concede. It is a bit odd to draw the battle lines in this way. But in doing so, we are only making things more difficult for ourselves. For we must argue that however the Practical Reasoning System works we do not predict and explain other people's behavior by taking the system off–line.

weak. Actual illustrations of such laws are sparse in number; and when examples are adduced, they commonly suffer from one of two defects: vagueness and inaccuracy.... But why, one wonders, should it be so difficult to articulate laws if we appeal to them all the time in our interpretative practice? [Goldman 1989, p. 167; see also Gordon 1986, p. 166 & Gordon forthcoming, a, Sec. 3.7]

Reply: Goldman is certainly right about one thing. It is indeed very difficult to articulate the principles of folk psychology precisely and accurately. But it is hard to see why this fact should be of any comfort to advocates of the off–line simulation theory. For much the same could be said about the knowledge structures underlying all sorts of cognitive capacities. It has proven enormously difficult to state the principles underlying a speaker's capacity to judge the grammaticality of sentences in his language. Indeed, after three decades of sustained effort, we don't have a good grammar for even a single natural language. Nor do we have a good account of the principles underlying people's everyday judgements about the behavior of middle sized physical objects, or about their ability to solve mathematical problems, or about their ability to play chess, etc. But, of course, in all of these domains, the theory–theory really is the only game in town. The off–line simulation story makes no sense as an account of our ability to judge grammaticality, or of our ability to predict the behavior of projectiles.

The difficulties encountered by those who have sought to describe the rules or principles underlying our grammatical (or mathematical or physical) abilities have convinced a growing number of theorists that our knowledge in these domains is not stored in the form of rules or principles. That conviction has been an important motive for the development of connectionist and other sorts of non–sentential and non–rule based models. But none of this should encourage an advocate of the off–line simulation theory. The dispute between connectionist models and rule–based models is the dispute between (C) and (D) in Figure 2.

And that is a dispute *among theory–theorists*. Of course on a narrow interpretation of "theory", on which only rule–

based and sentence–based models count as theories, the success of connectionism would indeed show that the "theory–theory" is mistaken. But, as we have taken pains to note, a refutation of the theory–theory will support the off–line simulation account only when "theory" is interpreted broadly.

Argument 2: Mental simulation models have been used with some success by a number of cognitive scientists.

Here's how Goldman makes the point:

> [S]everal cognitive scientists have recently endorsed the idea of mental simulation as one cognitive heuristic, although these researchers stress its use for knowledge in general, not specifically knowledge of others' mental states. Kahneman and Tversky 1982 propose that people often try to answer questions about the world by an operation that resembles the running of a simulation model. The starting conditions for a 'run', they say, can either be left at realistic default values or modified to assume some special contingency. Similarly, Rumelhart [et. al.] describe the importance of 'mental models' of the world, in particular, models that simulate how the world would respond to one's hypothetical actions. [Goldman 1989, p. 174]

Reply: Here, again, it is our suspicion that ambiguity between the two interpretations of "theory" is lurking in the background and leading to mischief. The "simulation" models that Goldman cites are the sort that would be classified under (D) in Figure 2. If they are used in the best explanation of a given cognitive capacity, then that capacity is subserved by a tacit theory, and *not* by an off–line simulation. Of course when "theory" is read narrowly, this sort of simulation will not count as a tacit theory. But, as already noted, on the narrow reading of "theory" the falsity of internalized theory accounts lends no support at all to the off–line simulation theory.

Argument 3: "To apply the alleged common–sense theory would demand anomalous precocity".

What we've just quoted is a section heading in one of Gordon's unpublished paper.[8] He goes on to note that recent

[8] Gordon (unpublished, a), Sec. 3.5.

studies have shown children as young as two and a half "already see behavior as dependent on belief and desire". It is, he suggests, more than a bit implausible that children this young could acquire and use "a theory as complex and sophisticated" as the one that the theory–theory attributes to them. Goldman elaborates the argument as follows:

> [C]hildren seem to display interpretive skills by the age of four, five or six. If interpretation is indeed guided by laws of folk psychology, the latter must be known (or believed) by this age. Are such children sophisticated enough to represent such principles? And how, exactly, would they acquire them? One possible mode of acquisition is cultural transmission (e.g. being taught them explicitly by their elders). This is clearly out of the question, though, since only philosophers have even tried to articulate the laws, and most children have no exposure to philosophers. Another possible mode of acquisition is private construction. Each child constructs the generalizations for herself, perhaps taking clues from verbal explanations of behavior which she hears. But if this construction is supposed to occur along the lines of familiar modes of scientific theory construction, some anomalous things must take place. For one thing, all children miraculously construct the same nomological principles. This is what the (folk–) theory theory ostensibly implies, since it imputes a single folk psychology to everyone. In normal cases of hypothesis construction, however, different scientists come up with different theories. [Goldman 1989, pp. 167-8]

Reply: There is no doubt that if the theory–theory is right, then the child's feat is indeed an impressive one. Moreover, it is implausible to suppose that the swift acquisition of folk psychology is subserved by the same learning mechanism that the child uses to learn history or chemistry or astronomy. But, once again, we find it hard to see how this can be taken as an argument against the theory–theory and in favor of the off–line simulation theory. For there are other cases in which the child's accomplishment is comparably impressive and comparably swift. If contemporary generative grammar is even *close* to being right, the knowledge structures that underlie a child's linguistic ability are enormously complex. Yet children seem to acquire the relevant knowledge structures even more quickly than they acquire their knowledge of folk psychology. Moreover, children in the same linguistic community all acquire much the same grammar, despite being exposed to significantly different samples of what will become their native language. Less is known about the knowledge structures underlying children's abilities to anticipate the behavior of middle sized physical objects. But there is every reason to suppose that this "folk physics" is at least as complex as folk psychology, and that it is acquired with comparable speed. Given the importance of all three knowledge domains, it is plausible to suppose that natural selection has provided the child with lots of help —either in the form of innate knowledge structures or in the form of special purpose learning mechanisms. But whatever the right story about acquisition turns out to be, it is perfectly clear that in the case of grammar, and in the case of folk physics, what is acquired must be some sort of internally represented theory. Off–line simulation could not possibly account for our skills in those domains. Since the speed of language acquisition and the complexity of the knowledge acquired do not (indeed, could not) support an off–line simulation account of linguistic ability, we fail to see why Gordon and Goldman think that considerations of speed and complexity lend any support at all to the off–line simulation account of our skills in predicting, explaining and interpreting behavior.

Argument 4: The off–line simulation theory is much simpler than the theory–theory.

Other things being equal, we should surely prefer a simpler theory to a more complex one. And on Gordon's view,

> the simulation alternative makes [the theory–theory] strikingly unparsimonious. Insofar as the store of causal generalizations posited by [the theory–theory] mirrors the set of rules *our own* thinking typically conforms to, the Simulation Theory renders it altogether otiose. For whatever rules our own thinking typically conforms to, our thinking continues to conform to them within the context of simulation.... In the light of this far simpler alternative, the hypothesis that people must be endowed with a special stock of laws corresponding to rules of logic and reason-

ing is unmotivated and unparsimonious. [Gordon forthcoming, a, Sec. 3, p. 7]

Reply: When comparing the simplicity of a pair of theories, it is important to look at the whole theory in both cases, not just at isolated parts. It is our contention that if one takes this broader perspective, the greater parsimony of the simulation theory simply disappears. To see the point, note that for both the theory–theory and the simulation theory the mechanism subserving our predictions of other people's behavior must have two components. One of these may be thought of as a data base that somehow stores or embodies information about how people behave. The other component is a mechanism which applies that information to the case at hand —it extracts the relevant facts from the data base. Now if we look only at the data base, it does indeed seem that the theory–theory is "strikingly unparsimonious" since it must posit an elaborate system of internally represented generalizations or rules —or perhaps some other format for encoding the regularities of folk psychology. The simulation theory, by contrast, uses the mind's decision making system as its "data base", and that decision making system would have to be there on any theory, because it explains how we make real, "on–line" decisions. So the off–line simulation theory gets its data base for free.

But now let's consider the other component of the competing theories. Merely *having* a decision making system will not enable us to make predictions about other people's behavior. We also need the capacity to take that system "off–line", feed it "pretend" inputs and interpret its outputs as predictions about how someone else would behave. When we add the required cognitive apparatus, the picture of the mind that emerges is sketched in Figure 3. Getting this "control mechanism" to work smoothly is sure to be a *very* non–trivial task. How do things look in the case of the theory–theory? Well, no matter how we go about making predictions about other people, it is clear that in making predictions about physical systems we can't use the off–line simulation strategy; we have to use some sort of internalized theory (though, of course, it need not be a sentence–like or rule–based theory). Thus we

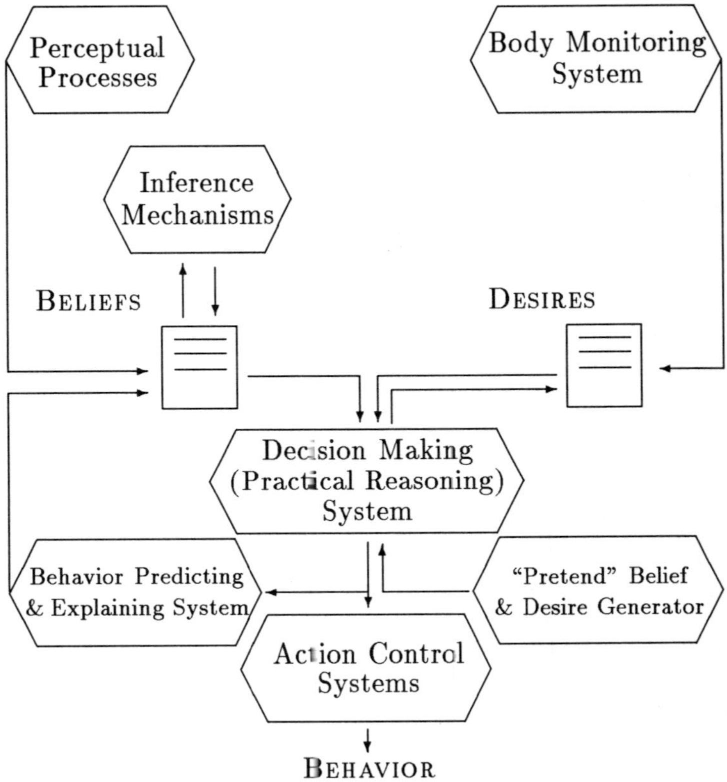

FIGURE 3.

know that the mind is going to have to have some mechanism for extracting information from internalized theories and applying it to particular cases. (In Figure 1 we have assumed that this mechanism is housed along with the other "Inference Mechanisms" that are used to extract information from pre–existing beliefs.) If such a mechanism will work for an internally represented folk physics, it is plausible to suppose that, with minor modifications, it will also work for an internally represented folk psychology. So while the simulation theorist gets the data base for free, it looks like the theory–theorist gets the "control mechanism" for free. All of this is a bit fast and loose, of course. But we don't think either side of this argument can get much more precise until we are presented with up–and–running models to compare. Until then, neither side can gain much advantage by appealing to simplicity.

Argument 5: When we introspect about our predictions of other people's behavior, it sometimes seems that we proceed by imagining how we would behave in their situation.

Here is how Goldman makes the point:

> The simulation idea has obvious initial attractions. Introspectively, it seems as if we often try to predict others' behavior —or predict their (mental) choices— by imagining ourselves in their shoes and determining what we would choose to do. [Goldman 1989, p. 169]

And here is Gordon:

> [C]hess players report that, playing against a human opponent or even against a computer, they visualize the board from the other side, taking the opposing pieces for their own and vice versa. Further, they pretend that their reasons for action have shifted accordingly.... Thus transported in imagination, they 'make up their mind what to do.' [Gordon 1986, pp. 161-2]

Both authors are aware that appeal to introspection can be a two edged sword, since it also often happens that we predict other people's behavior *without* introspecting any imaginary behavior.

> [T]here is a straightforward challenge to the psychological plausibility of the simulation approach. It is far from obvious, introspectively, that we regularly place ourselves in another person's shoes, and vividly envision what we would do in his circumstances. [Goldman 1989, p. 176]
>
> Imagery is not always needed in such simulations. For example, I need no imagery to simulate having a million dollars in the bank. [Gordon 1986, p. 161]

To deal with this "challenge", Goldman proposes a pair of replies. First, simulation need not always be introspectively vivid. It can often be "semi–automatic, with relatively little salient phenomenology". [Goldman 1986, p. 176] Second, not all interpretations rely on simulation. In many cases interpreters rely solely on "inductively acquired information" though the information is "historically derived from earlier simulations". [Goldman 1986, p. 176]

Reply: We don't propose to make any fuss at all about the frequent absence of "salient phenomenology". For it is our contention that when the issue at hand is the nature of the cognitive mechanism subserving our capacity to interpret and predict other people's behavior, the entire issue of introspective imagination is a red herring. Indeed, it is *two* red herrings. To see the first of them, consider one of the standard examples used to illustrate the role of imagery in thought. Suppose we ask you: "How many windows are there in your house?" How do you go about answering? Almost everyone reports that they *imagine* themselves walking from room to room, counting the windows as they go. What follows from this about the cognitive mechanism that they are exploiting? Well, one thing that surely *does not* follow is that off–line simulation is involved. The *only* way that people could possibly answer the question accurately is to tap into some internally represented store of knowledge about their house; it simply makes no sense to suppose that off–line simulation is being used here. So even if a cognitive process is *always* accompanied by vivid imagery, that is no reason at all to suppose that the process exploits off–line simulation. From this we draw the obvious conclusion. The fact that prediction and interpretation *sometimes* involve imagining oneself in the other person's shoes is less than no reason at all to suppose that off–line simulation is involved.

It might be suggested that, though imagery provides no support for the off–line simulation hypothesis, it does challenge the theory–theory when "theory" is interpreted narrowly. For it shows that some of the information we are exploiting in interpretation and prediction is not stored in the form of sentences or rules. But even this is far from obvious. There is a lively debate in the imagery literature in which "descriptionalists", like Pylyshyn and Dennett, maintain that the mechanisms underlying mental imagery exploit language–like representations, while "pictorialists", like Kosslyn and Fodor, argue that images are subserved by a separate, non–linguistic sort of representation. [Pylyshyn 1981,

Dennett 1969 & 1978, Fodor 1975, Kosslyn 1981] We don't propose to take sides in this dispute. For present purposes it is sufficient to note that, unless it is supplemented by a persuasive argument in favor of pictorialism and against descriptionism, the introspective evidence does not even challenge the theory–theory *construed narrowly.*

Argument 6: The off–line simulation account is supported by recent experimental studies focusing on children's acquisition of the ability to interpret and predict other people.

On our view, this is far and away the most interesting argument that has been offered in favor of the off–line simulation theory. To see exactly what the experimental studies do, and do not, support, we'll have to look at both the evidence and the argument with considerable care. Gordon does a good job of describing one important set of experiments.

> Very young children give verbal expression to predictions and explanations of the behavior of others. Yet up to about the age of four they evidently lack the concept of belief, or at least the capacity to make allowances for false or differing beliefs. Evidence of this can be teased out by presenting children with stories and dramatizations that involve dramatic irony: where we the audience know something important the protagonist doesn't know. . . .
>
> In one such story (illustrated with puppets) the puppet–child Maxi puts his chocolate in the box and goes out to play. While he is out, his mother transfers the chocolate to the cupboard. Where will Maxi look for the chocolate when he comes back? In the box, says the five year old, pointing to the miniature box on the puppet stage: a good prediction of a sort we ordinarily take for granted. . . . But the child of three to four years has a different response: verbally or by pointing, the child indicates the cupboard. (That is, after all, where the chocolate is to be found, isn't it?) Suppose Maxi wants to mislead his gluttonous big brother to the *wrong* place, where will he lead him? The five year old indicates the cupboard, where (unbeknownst to Maxi) the chocolate actually is. . . . The *younger* child indicates, incorrectly, the box. [Gordon 1986, p. 168]

These results, Gordon maintains, are hard to square with the theory–theory. For if the theory–theory is correct, then

> before internalizing [the laws and generalizations of folk psychology] the child would simply be unable to predict or explain human action. And *after* internalizing the system, the child could deal indifferently with actions caused by *true* beliefs and actions caused by *false* beliefs. It is hard to see how the semantical question could be relevant. [Gordon 1986, p. 169]

But, according to Gordon, these data are just what we should expect, if the off–line simulation theory is correct.

> The Simulation Theory [predicts that] prior to developing the capacity to simulate others for purposes of prediction and explanation, a child will make *egocentric errors* in predicting and explaining the actions of others. She will predict and explain as if whatever she herself counts as "fact" were also fact to the other; which is to say, she fails to make allowances in her predictions and explanations for false beliefs or for what the other isn't in a position to know. [Gordon unpublished, a, Sec. 3.6, p. 11]

Reply: According to Gordon, the theory–theory can't easily explain the results of the "Maxi" experiment, though the off–line simulation theory predicts those results. We're not convinced on either score. Let's look first at just what the off–line simulation story would lead us to expect.

Presumably by the time any of these experiments can be conducted, the child has developed a more or less intact decision making system like the one depicted in Figure 1. That system makes "on–line" decisions and thus determines the child's actions on the basis of her actual beliefs and desires. But by itself it provides the child with no way of predicting Maxi's behavior or anyone else's. If the off–line simulation theory is right, then in order to make predictions about other people's behavior two things must happen. First, the child must acquire the ability to take the output of the decision making system off–line —treating its decisions as predictions or expectations, rather than simply feeding them into the action controlling system. Second, the child must acquire the ability to provide the system with input other than her own actual beliefs and desires. She must be able to supply the system with "pretend" input so that she can predict the behavior of someone whose beliefs and desires are different from her own. (These are the two capacities that are represented in Figure 3 and absent in Figure 1.) There is, of course, no *a priori* reason to suppose that these two steps happen at different times, nor that the one we've listed first will occur first. But if they do occur in that order, then we might expect there to be a period when the child could predict her own behavior (or the behavior of someone whose beliefs and desires are the same as hers) though she could not predict the behavior of people whose beliefs or desires are different from hers. It is less clear what to expect if the steps occur in the opposite order. Perhaps the result would be some sort of pretending or play–acting —behaving in a way that someone with different beliefs or desires would behave. Though until the child develops the capacity to take output of the decision making system off–line, she will not be able to predict other people's behavior or her own. So it looks like the off–line simulation story makes room for three possible developmental scenarios.

(1) The child acquires both abilities at the same time. In this case we would expect to see two developmental stages. In the first the child can make no predictions. In the second she can make a full range of predictions about people whose beliefs and desires are different from her own.

(2) The child first acquires the ability to take the output off–line, and then acquires the ability to provide the system with pretend input. In this case we would expect three developmental stages. In the first, the child can make no predictions. In the second, she can only make predictions about her own behavior or about the behavior of people whose beliefs and desires are identical to hers. In the third, she can make the full range of predictions.

(3) The child first acquires the ability to provide the system with pretend inputs, and then acquires the ability to take the output off–line. In this case, too, we would expect three developmental stages. The first and last

stages are the same as those in (2), but in the middle stage the child can play–act but not make predictions.

Now let's return to the Maxi experiment. Which of these developmental scenarios do the children in these experiments exhibit? At first blush, it might be thought that the pattern Gordon reports is much the same as the one set out in scenario (2). But that would be a mistake. The younger children —those who are giving the wrong answers— are not predicting that Maxi would do what someone with their own beliefs and desires would do. For they have no desire to get the chocolate, nor to deceive the gluttonous brother. Those are *Maxi's* desires, not *theirs*. If anything, it would appear that these children are half way between the second and third stages of scenario (2): they can feed "pretend" desires into the decision making system, but not "pretend" beliefs. Of course none of this shows that the off–line simulation theory is false. It is perfectly compatible with the theory to suppose that development proceeds as in (2), *and* that the transition from the second to the third stage proceeds in two sub–stages —desires first, and then beliefs. (This pattern is sketched in Figure 4.) But it is, to say the least, something of an exaggeration to say that the off–line simulation theory "predicts" the experimental results. The most that can be said is that the theory is compatible with the observed developmental pattern, and with lots of other patterns as well.[9]

For the results that Gordon describes to be at all relevant to the dispute between the off–line simulation theory and the theory–theory it would have to be the case that the latter theory is *in* compatible with the reported developmental pattern. But that is patently not the case. To see why, we should first note that the theory–theory is not committed to the claim that folk psychology is acquired all in one

[9] Actually, the developmental facts are rather more complicated than Gordon suggests. For, as Leslie (1988) emphasizes, children are typically able to appreciate and engage in pretend play by the time they are two and a half years old —long before they can handle questions about Maxi and his false beliefs. It is not at all clear how the off–line simulation theory can explain both the early appearance of the ability to pretend and the relatively late appearance of the ability to predict the behavior of people whose beliefs and desires differ from one's own.

fell swoop. Indeed, one would expect just the opposite. If children really are acquiring a tacit theory of the mind, they probably acquire it a bit at a time. Thus it might be the case that, at a given stage in development, children have mastered the part of the theory that specifies how beliefs and desires lead to behavior, though they have not mastered the entire story about how beliefs are caused. At this stage, they might simply assume that beliefs are caused by the way the world is; they might adopt the strategy of attributing to everyone the very same beliefs that they have. A child who has acquired this much of folk psychology would (incorrectly) attribute to Maxi the belief that the chocolate is in the cupboard. She would then go on to make just the predictions that Gordon reports. Of course, the theory–theory is also compatible with lots of other hypotheses about which bits of folk psychology are acquired first. Thus, like the off–line simulation theory, it is compatible with (but does not entail) lots of possible developmental patterns. So it looks like the developmental studies that Gordon and Goldman cite can't be used to support one theory over the other.

Argument 7: Autistic children are highly deficient in their ability to engage in pretend play. These children are also frequently unable to impute beliefs to others or to predict other people's behavior correctly.

Here's how Gordon sets out the argument:

> Practical simulation involves the capacity for a certain kind of systematic pretending. It is well known that *autistic* children suffer a striking deficit in the capacity for pretend–play. In addition, they are often said to 'treat people and objects alike'; they fail to treat others as subjects, as having 'points of view' distinct from their own. This failure is confirmed by their performance in prediction tests like the [Wimmer–Perner "Maxi" experiment] I have just described. A version of the Wimmer–Perner test was administered to autistic children of ages *six to sixteen* by a team of psychologists.... *Almost all* these children gave the wrong answer, the 3–year–old's answer. This indicates a highly specific deficit, not one in general intelligence. Although many autistic children are also mentally retarded, those tested were mostly in the average or borderline IQ range. Yet children with Down's

> syndrome, with IQ levels substantially below that range, suffered no deficit: almost all gave the right answer. My account of belief would predict that only those children who can engage in pretend play can master the concept of belief. [Gordon 1986, p. 196]

Goldman is rather more tentative. He claims only that the inability of autistic children "to impute beliefs to others and therefore predict their behavior correctly...might...be related to their lack of pretend play". [Goldman 1989, p. 175]

Reply: The fact that autistic children are both incapable of pretend play and unable to predict the behavior of other people in Wimmer–Perner tests is very intriguing. Moreover, Gordon is certainly right in suggesting that the off–line simulation theory provides a possible explanation for these facts. If the off–line simulation theory is right, predicting the behavior of people whose beliefs differ from our own requires an ability to provide our own decision making system with pretend input. And it is plausible to assume that this ability would also play a central role in pretend play. So if we hypothesize that autistic children lack the ability to provide the decision making system with pretend input, we could explain both their performance on the Wimmer–Perner test and their failure to engage in pretend play. But, of course, this will not count as an argument for the off–line simulation theory and against the theory–theory if the latter account can offer an equally plausible explanation of the facts. And it will require no creativity on our part to produce such an alternative explanation since one of the investigators who discovered the fact that autistic children do poorly in Wimmer–Perner tests has offered one himself.

Leslie (1988) takes as an assumption "the hypothesis that human cognition involves *symbolic computations* in the sense discussed...by Newell (1980) and particularly by Fodor". [Leslie 1988, p. 21] He also assumes that an internalized theory of mind underlies the normal adult's ability to predict other people's behavior. An important theme in Leslie's work is that developmental studies with both normal and autistic children can help to illuminate the expressive resources of the "language of thought" in which our theory of mind is encoded. According to Leslie, the notion of a "meta–

representation" is central in understanding how our theory of mind develops. Roughly speaking, a meta–representation is a mental representation about some other representational state or process. We exploit meta–representations when we think that

Maxi believes that the chocolate is in the box

or that

Maxi's brother wants the chocolate

or that

Mommy is pretending that the banana is a telephone.

On Leslie's view, "autistic children do not develop a theory of mind normally". [Leslie 1988, p. 39] And while "it is far too soon to say with any confidence what *is* wrong" with these children, he speculates that at the root of the problem may be an inability to use meta–representations. If this were true, it would explain both their difficulty with pretend play and their failure on the Wimmer–Perner test.

Though we find Leslie's speculation interesting and important, it is no part of our current project to defend it. To make our case we need only insist that, on currently available evidence, Leslie's hypothesis is no less plausible than Gordon's. Since Leslie's speculation presupposes that normal children acquire and exploit a theory of mind that is encoded in a language of thought, the evidence from studies of autistic children gives us no reason to prefer the off–line simulation account over the theory–theory.[10]

---

[10] It's worth noting that both Gordon's account and Leslie's "predict that only those children who can engage in pretend play can master the concept of belief". [Gordon 1986, p. 169] This may prove a troublesome implication for both theorists, however. For it is not the case that *all* autistic children do poorly on the Wimmer–Perner test. In the original study reported by Baron–Cohen, Leslie and Frith (1985), only 16 out of 20 autistic subjects failed the Wimmer–Perner test. The other 4 answered correctly. The investigators predicted that these children "would also show evidence of an ability to pretend play". (p. 43) Unfortunately, no data was reported on the pretend play ability of these

Our theme, in this reply and in the previous one, has been that the empirical evidence cited by Gordon and Goldman, while compatible with the off-line simulation theory, is also compatible with the theory-theory, and thus does not support one theory over the other. But there are other studies in the recent literature that *can* be used to support one theory over the other. These studies report results that are comfortably compatible with the theory-theory though not with the off-line simulation account. Before we sketch those results, however, it is time to start a new section. In this section we've tried to show that none of the arguments in favor of the off-line simulation theory are persuasive. In the next one we'll set out a positive case for the theory-theory.

5 In Defense of the Theory-Theory

Argument 1: There are developmental data that are easily accommodated by the theory-theory, but very hard to explain if the off-line simulation account is correct.

Let's start with a description of the experimental setup, and a quick overview of the data.

> The setup of the task in these experiments was rather simple. Two children were placed facing each other on opposite sides of a table. In each trial one child served as subject and had access to the other child's knowledge and his or her own knowledge of the content of a closed box. The box was placed in the middle of the table between the two children. The outside of it was neutral and not suggestive of its content. In each box was a familiar object like a pencil, a comb, a piece of chocolate, and so on. The specific questions were: "Does (name of other child)

subjects. If it should turn out that some autistic children do well on the Wimmer-Perner test *and* lack the ability for pretend play, both Gordon's explanation and Leslie's would be in trouble. If the facts do turn out this way, advocates of the theory-theory will have a variety of other explanations available. But it is much less clear that the off-line simulation account could explain the data, if some autistic children can't pretend but can predict the behavior of people with false beliefs.

> know what is in the box or does she not know that?" and "Do you know what is in the box or do you not know that?"....
>
> Before the knowledge–questions were asked, either the other or the subject had access to the content of the box. One kind of access was visual perception. In this case either the other child or the subject had a chance to look into the box. The other kind of access was verbal information. Here the experimenter looked into the box and then informed one of the children by whispering the name of the content object into the child's ear. Because the two children were facing each other the subject was fully aware of the informational conditions the other child was exposed to, that is, of whether the other child did or did not look into the box and of whether the other was or was not informed. [Wimmer, et. al. 1988, p. 175]

The results of this experiment were quite striking. The older children (5–year–olds) gave uniformly correct answers. But younger children (3– and 4–year–olds) did not.

> The most frequent error was denial of the other child's knowledge when the other child had looked into the box or was informed by the experimenter.
>
> Most 3–year–olds and some 4–year–olds said that the other did not know what was in the box. This kind of error was nearly absent in children's assessment of their own knowledge. When subjects themselves had looked into the box or were informed, then they claimed to know and they could, of course, tell what was in the box. [Wimmer, et. al. 1988, pp. 175-6]

In another experiment, designed to be sure that the younger children were aware the other child had looked in the box, the subjects were asked both whether the other child had looked in the box and whether the other child knew what was in the box. "The children consistently responded affirmatively to the look–question but again quite frequently responded negatively to the knowledge question". [Wimmer, et. al. 1988, p. 176]

What is going on here? The explanation offered by the experimenters is that younger children are using quite different mental processes in assessing what they know and in assessing what the other child knows. To answer the question. "Do you know what is in the box?" the children use what the experimenters call the "answer check procedure". They simply check to see whether they have an answer to the embedded question in their knowledge base, and if they do they respond affirmatively. To answer the question about the other child's knowledge, the older children used what the experimenters call a "direct access check procedure". In effect, they ask themselves whether the other child looked in the box or was told about its contents. If so, they respond affirmatively. If not, they respond negatively. However, the 3–year–olds did not use this procedure. They simply checked whether the other child had uttered a correct statement about the box's contents. If she had not, the subject said the other did not know. A very natural way to describe the situation is that while the younger children know that people who say *that p* typically believe or know *that p*, these children have not yet learned that people will come to know *that p* by seeing or being told *that p*. The younger children have acquired a fragment of folk psychology, while the older children have acquired a more substantial piece of the theory.[11] The older

[11] Another experiment reported by Perner et al. (1987) provides some additional evidence for this conclusion. In the first part of the experiment children were shown a box of Smarties (a type of candy), and asked what they thought was in the box. All of them answered that the box contained Smarties. They were then shown that the box contained a pencil, and no Smarties. After this the children were asked three questions: i) What is in the box. ii) What did you think was in the box when you first saw it. iii) What would a friend, waiting outside, think was in the box if he saw it as it is now.

Most of the younger children answered (iii) incorrectly; they failed to predict their friend's false belief. But more than half of those who got (iii) wrong answered (ii) correctly. They were able to tell the experimenter that they had thought the box contained Smarties, and that they were wrong. In commenting on this experiment, Leslie notes that

> [d]espite the ability to report their false belief, these 3–year–olds could not understand where that false belief had come from.... Despite the fact that they themselves had just undergone the process of getting that false belief, the children were quite unable to understand and reconstruct that process, and thus unable, minutes later, to predict what would happen to their friend. [Leslie 1988, pp. 33-4.]

children have not, however, entirely mastered the theory, as indicated by another series of experiments.

These experiments focused on the role of *inference* in the acquisition of knowledge or belief. What they show is that "four– and 5–year olds relied on inference in their own acquisition of knowledge but denied that the other person might know via inference". [Wimmer, et. al. 1988, p. 179]

> Inferential access was realized in these experiments in a very simple and concrete way. In a first step the child and the other person together inspected the content of a container and agreed that only sweets of a certain kind, for example, black chocolate nuts, were in the container. In a second step the other person or the subject was prevented from seeing how one choconut was transferred from the container into an opaque bag. However, this person was explicitly informed by the experimenter about this transfer, for example, "I've just taken one of the things out of this box and put it in the bag".
>
> The condition where knowledge could be acquired via simple inference was contrasted with a condition where knowledge depended on actually seeing the critical object's transfer. In this latter condition two kinds of sweets were in the original container, and thus one could only know what the content of the critical bag was by having seen the transfer from container to bag. [Wimmer, et. al. 1988, p. 179]

Once again, the results were quite striking. In most cases the older children (6–year–olds, in this case) generally gave the right answers both about their own knowledge and about the other child's knowledge. But although the 4–year–olds used inference in forming their own beliefs, a substantial majority of them exhibited a pattern that the experimenters called "inference neglect".

> The response pattern "inference neglect" means that the other person was assessed according to perceptual access: When the other person saw the object's transfer to the bag, 4–year–olds attributed knowledge; when the other did not see this transfer, ignorance was attributed even when the other person in fact knew via inference. [Wimmer, et. al. 1988, p. 179]

One plausible way of accounting for these results is to hypothesize that the older children had mastered yet another part of the adult folk psychology. They had learned that knowledge and beliefs can be caused by inference as well as by direct perceptual access. And, indeed, this is just the interpretation that the experimenters suggest.

> In contrast to the 3–year–olds discussed in the previous [experiment], the 4–and 5–year–olds in the present experiments understood quite well that one has to consider the other person's informational conditions when one is questioned about the other person's knowledge. Their only problem was their limited understanding of informational conditions. They understood only direct visual access as a source of knowledge and this led them to mistaken but systematic ignorance attributions in the case of inferential access. [Wimmer, et. al. 1988, p. 181]

Let's now ask what conclusions can be drawn from these experiments that will be relevant to the choice between the off–line simulation theory and the theory–theory. A first obvious fact is that the data are all comfortably compatible with the theory–theory. Indeed, the explanation of the data offered by the experimenters is one that presupposes the correctness of the theory–theory. What appears to be happening is that as children get older, they master more and more of the principles of folk psychology. By itself, of course, the theory–theory would not enable us to predict the data, since the theory–theory does not tell us anything about the order in which the principles of folk psychology are acquired. But the pattern of results described certainly poses no problem for the theory–theory.

The same cannot be said for the off–line simulation theory. It is clear that even the younger children in these studies form beliefs as the result of perception, verbally provided information, and inference. So there is nothing about their decision making system, when it is being used on–line, that will help to explain the results. To make predictions about other people, the off–line simulation theory maintains, children must acquire the capacity to take the decision making system off–line and provide it with some pretend inputs. But there is no obvious way in which this process could produce the pattern of results that has been reported. The difficulty is particularly clear in the case of inference. If the subject has seen that the box contains only chocolate nuts, and if she is told that one of the items in the box has been put in the bag, she comes to believe that there is a chocolate nut in the bag. But if she knows the other child has also seen what is in the box, and that the other child has been told that one of the items in the box has been put in the bag, she insists that the other child does not know what is in the bag. The problem can't be that the subject doesn't realize that the other child knows what is in the box. Children of this age do a good job of attributing belief on the basis of perception. Nor can it be that the subject doesn't believe that the other child believes the transfer has been made. For children of this age are also adept at attributing beliefs on the basis of verbally communicated information. So it looks like the subject has all the information needed for a successful simulation. But the answer she comes up with is *not* the one that she herself would come up with, were she in the subject's place. There are, of course, endlessly many ways in which a resolute defender of the off–line simulation theory might try to accommodate these data. But all the ones we've been able to think of are obviously implausible and ad hoc.

Argument 2: Our predictions and explanations of behavior are "cognitively penetrable".

One virtue of using a simulation to predict the behavior of a system is that you need have no serious idea about the principles governing the behavior of the target system. You just run the simulation and watch what happens. Sometimes, of course, a simulation will do something that was utterly unexpected. But no matter. If the simulation really was similar to the target system, then the prediction it provides will be a good one. In predictions based on simulations, what you don't know won't hurt you. All of this applies to the off–line simulation theory, of course. If there is some quirk in the human decision making system, something quite unknown to most people that leads the system to behave in an unexpected way under certain circumstances, the accuracy of predictions based on simulations should not be adversely affected. If you provide the system with the right pretend input, it should simulate (and thus predict) the unexpected output. Adapting a term from Pylyshyn, we might describe this by saying that simulation–based predictions are not "cognitively penetrable".[12]

Just the opposite is true for predictions that rely on a theory. If we are making predictions on the basis of a set of laws or principles, and if there are some unexpected aspects of the system's behavior that are not captured by our principles, then our predictions about those aspects of the system's behavior should be less accurate. Theory based predictions are sensitive to what we know and don't know about the laws that govern the system; they *are* cognitively penetrable. This contrast provides a useful way of testing the two theories. If we can find cases in which ignorance about the workings of one's own psychology leads people to make mistakes in predicting what they, or other similarly situated people, would do, it will provide yet another reason to think that the off–line simulation theory is untenable. And, as it happens, cases illustrating cognitive penetrability in the prediction of behavior are not all that hard to find. The literature in cognitive social psychology is full of them. We'll illustrate the point with three examples, but it would be easy to add three dozen more.

First Example: Suppose you are walking through the local shopping mall, and encounter what looks to be yet another consumer product opinion survey. In this one a polite, well dressed man invites you to examine an array of familiar products —nightgowns, perhaps, or pantyhose— and to rate their quality. A small reward is offered for your participation — you can keep the garment you select. On examining the products, you find no really significant differences among them. (You couldn't, because, unbeknownst to you they are identical.) What would you do? Confronted with this question, most of us think we would report that the garments looked to be very similar, and then choose one randomly. However, when the experiment was actually tried, this turned out to be mistaken. "There was a pronounced position effect on

[12] Pylyshyn 1981 & 1984. It is perhaps worth noting that we are using the term "cognitively penetrable" a bit more loosely than Pylyshyn does. But in the present context the difference is not important.

evaluations, such that the right–most garments were heavily preferred to the left–most garments". But it was clear that few of the subjects had any awareness at all of the effect of position on their decision. Indeed, "when questioned about the effect of the garments' position on their choices, virtually all subjects denied such an influence (usually with a tone of annoyance or of concern for the experimenter's sanity)". [Nisbett & Ross 1980, p. 207]

This sort of case poses real problems for the off–line simulation theory. Most people have no trouble imagining themselves in the situation described. They can supply their decision making system with vivid "pretend" input. But few people who have not heard of the experiment predict that they would behave in the way that the subjects behaved. The natural interpretation of the experiment is that people's predictions about their own behavior (and the subjects' explanations of their own choice) are guided by an incomplete or inaccurate theory, one which includes no information about these so–called "position effects".

Second Example: Here's another case to run through your own simulator. Suppose someone in the office is selling $1.00 tickets for the office lottery. In some cases, when a person agrees to buy a ticket, he or she is simply handed one. In other cases, after agreeing to buy a ticket, the buyer is allowed to choose a ticket from several that the seller has available. On the morning of the lottery, the seller approaches each purchaser and attempts to buy back the ticket. Now imagine yourself in both roles —first as a person who had been handed the ticket, second as a person who had been given a choice. What price would you ask in each case? Would there be any difference between the two cases? On several occasions one of us (Stich) has asked large undergraduate classes to predict what they would do. Almost no one predicts that they would behave the way that people actually do behave. Almost everyone is surprised to hear the actual results.

Ah, yes, the results; we haven't yet told *you* what they are. When the experiment was actually done, "no–choice subjects sold their tickets back for an average of $1.96. Choice subjects, who had personally selected their tickets, held out for an average of $8.67!" [Nisbett & Ross 1980, p. 136] If, like Stich's students, you find this surprising and unexpected, it counts as yet another difficulty for the off–line simulation theory.

Third Example: In the psychology laboratory, and in everyday life, it sometimes happens that people are presented with fairly persuasive evidence that they have some hitherto unexpected trait. In the light of that evidence people form the belief that they have the trait. What will happen to that belief if, shortly after this, people are presented with a convincing case discrediting the first body of evidence? Suppose, for example, they are convinced that the test results were actually someone else's, or that no real test was conducted at all. Most people expect that the undermined belief will simply be discarded. If until recently I never had reason to think I had a certain trait, and if the evidence I just acquired has been soundly discredited, then surely it would be silly of me to go away thinking that I *do* have the trait. That seems to be what most people think. And the view was shared by a generation of social psychologists who duped subjects into believing all sorts of things about themselves, observed their reactions, and then "debriefed" the subjects by explaining the ruse. The assumption was that no enduring harm could be done because once the ruse was explained the induced belief would be discarded. But in a widely discussed series of experiments, Ross and his co–workers have demonstrated that this is simply not the case. Once a subject has been convinced that she is very good at telling real from fake suicide notes, for example, showing her that the evidence was completely phony does not succeed in eliminating the belief. Moreover, third person observers of the experiment exhibit even stronger "belief perseverance". If an observer subject watches a participant subject being duped and then debriefed, the observer, too, will continue to believe that the participant is particularly good at detecting real suicide notes. [Nisbett & Ross 1980, p. 175-179]

Neither of these results should have been at all surprising to anyone if we predict each other's beliefs and behavior in the way that the off–line simulation theory suggests. But clearly the results *were* both surprising and disturbing. We can't simply ask ourselves what we would do in these circumstances and expect to come up with the right answer. For the theory–theorist, this fact poses no particular problem. When our folk psychology is wrong, it is to be expected that our predictions will be wrong too. It is simply another illustration of cognitive penetrability in predicting and explaining behavior. The theory–theory, unlike the off–line simulation theory, predicts that people's predictions and explanations of behavior will be cognitively penetrable through and through. If it is agreed that these experiments confirm cognitive penetrability, the off–line simulation theory is in serious trouble.

6 Conclusion

Our paper has been long but our conclusion will be brief. The off–line simulation theory poses an intriguing challenge to the dominant paradigm in contemporary cognitive science. Moreover, if it were correct the off–line simulation account of psychological prediction and explanation would largely undermine *both* sides in the eliminativism dispute. But it has been our contention that the prospects for the off–line simulation theory are not very bright. None of the arguments that have been offered in defense of the theory are at all persuasive. And there is lots of experimental evidence that would be very hard to explain if the off–line simulation account were correct. We don't claim to have provided a knock–down refutation of the off–line simulation theory. Knock–down arguments are hard to come by in cognitive science. But we do claim to have assembled a pretty serious case against the simulation theory. Pending a detailed response, we don't think the off–line simulation theory is one that cognitive scientists or philosophers should take seriously.[13]

---

[13] We are grateful to Jerry Fodor for his helpful comments on an earlier version of this paper. Thanks are also due to Joseph Franchi for help in preparing the Figures.

7 REFERENCES

Astington, J., Harris, P., and Olson, D. (eds.) 1988. *Developing Theories of Mind.* Cambridge: Cambridge University Press.

Baron–Cohen, S., Leslie, A. and Frith, U. 1985. Does the Autistic Child Have a "Theory of Mind"? *Cognition*, 21, 37-46.

Chomsky, N. 1965. *Aspects of the Theory of Syntax.* Cambridge, MA: MIT Press.

Churchland, P. 1981. Eliminative Materialism and the Propositional Attitudes. *Journal of Philosophy*, 78, 67-90.

Churchland, P. 1989. Folk Psychology and the Explanation of Human Behavior. In Churchland, *A Neurocomputational Perspective.* Cambridge, MA: MIT Press. Pp. 111-128.

Cummins, R. 1983. *The Nature of Psychological Explanation.* Cambridge, MA: MIT Press.

D'Andrade, R. 1987. A Folk Model of the Mind. In Holland, D., and Quinn, N. (eds.), *Cultural Models in Language and Thought.* Cambridge: Cambridge University Press.

Dennett, D. 1969. *Content and Consciousness.* London: Routledge and Kegan Paul.

Dennett, D. 1978a. Artificial Intelligence as Philosophy and Psychology. In Dennett, *Brainstorms.* Cambridge, MA: MIT Press. Pp. 109-126.

Dennett, D. 1978b. Two Approaches to Mental Images. In Dennett, *Brainstorms.* Cambridge, MA: MIT Press. Pp. 174-189.

Fodor, J. 1987. *Psychosemantics.* Cambridge, MA: MIT Press.

Fodor, J. 1968: The Appeal to Tacit Knowledge in Psychological Explanation. *Journal of Philosophy*, 65, 000-00. Reprinted in Fodor 1981.

Fodor, J. 1975. *The Language of Thought.* New York: Thomas Crowell.

Fodor, J. 1981: *Representations.* Cambridge, MA: MIT Press.

Fodor, J., Bever, T. and Garrett, M. 1974. *The Psychology of Language: An Introduction to Psycholinguistics and Generative Grammar.* New York: McGraw–Hill.

Goldman, A. 1989: Interpretation Psychologized. *Mind and Language*, 4, 161-185.

Gordon, R. 1986: Folk Psychology as Simulation. *Mind and Language*, 1, 158-171.

Gordon, R. unpublished, a: Fodor's Intentional Realism and the Simulation Theory. MS dated 2/90.

Gordon, R. unpublished, b: Simulation and the Theory–Theory. MS dated 1/25/91. Presented at Pacific APA, 1991.

Gregory, R. 1970. *The Intelligent Eye*. New York: McGraw–Hill.

Greno, J. 1983. Conceptual Entities. In Gentner, D., and Stevens, A. (eds.) *Mental Models*. Hillsdale, NJ: Erlbaum. Pp. 227-252.

Hayes, P. 1985. The Second Naive Physics Manifesto. In Hobbs, J., and Moore, R. (eds.), *Formal Theories of the Commonsense World*. Norwood, NJ: Ablex. Pp. 1-36.

Heal, J. 1986: Replication and Functionalism. In J. Butterfield (ed.), *Language, Mind and Logic*. Cambridge: Cambridge University Press. Pp. 135-150.

Johnson–Laird, P. 1983. *Mental Models: Towards a Cognitive Science of Language, Inference and Consciousness*. Cambridge, MA: Harvard University Press.

Kahneman, D. and Tversky, A. 1982. The Simulation Heuristic. In Kahneman, D., Slovic, P. and Tversky, A. (eds.) *Judgment Under Uncertainty*. Cambridge: Cambridge University Press.

Kosslyn, S. 1981. The Medium and the Message in Mental Imagery: A Theory. In Block, N. (ed.), *Imagery*. Cambridge, MA: MIT Press. Pp. 207-244.

Leslie, A. 1987. Pretense and Representation: The Origins of "Theory of Mind". *Psychological Review*, 94, 412-426.

Leslie, A. 1988. Some Implications of Pretense for Mechanisms Underlying the Child's Theory of Mind. In Astington, J., Harris, P. and Olson, D. (eds.), *Developing Theories of Mind*. Cambridge: Cambridge University Press. Pp. 19-46.

Lycan, W. 1981. Form, Function and Feel. *The Journal of Philosophy*, 78, 24-50.

Lycan, W. 1988. Toward a Homuncular Theory of Believing. In Lycan, *Judgement and Justification*. Cambridge: Cambridge University Press.

Marr, D. 1982. *Vision*. San Francisco: Freeman.

McCloskey, M. 1983. Naive Theories of Motion. In Gentner, D. and Stevens, A. (eds.), *Mental Models*. Hillsdale, NJ: Erlbaum.

Montgomery, R. 1987: Psychologism, Folk Psychology and One's Own Case. *Journal for the Theory of Social Behavior*, 17, 195-218.

Newell, A. 1980. Physical Symbol Systems. *Cognitive Science*, 4, 135-183.

Nisbett, R. and Ross, L. 1980. *Human Inference*. Englewood Cliffs, NJ: Prentice–Hall.

Olson, D., Astington, J. and Harris, P. 1988. Introduction. In Astington, J., Harris, P., and Olson, D. (eds.) 1988. *Developing Theories of Mind*. Cambridge: Cambridge University Press.

Perner, J., Leekam, S. and Wimmer, H. 1987. Three–year–olds' Difficulty with False Belief: The Case for a Conceptual Deficit. *British Journal of Developmental Psychology*, 5, 125-137.

Pinker, S. 1989. *Learnability and Cognition*. Cambridge, MA: MIT Press.

Pylyshyn, Z. 1981. The Imagery Debate: Analog Media versus Tacit Knowledge. In Block, N. (ed.), *Imagery*. Cambridge, MA: MIT Press. Pp. 151-206.

Pylyshyn, Z. 1984. *Computation and Cognition*. Cambridge, MA: MIT Press.

Ramsey, W., Stich, S. and Garon, J. 1990. Connectionism, Eliminativism and the Future of Folk Psychology. *Philosophical Perspectives*, 4, 499-533.

Ripstein, A. 1987: Explanation and Empathy. *Review of Metaphysics*, 40, 465-482.

Rock, I. 1983. *The Logic of Perception*. Cambridge, MA: MIT Press.

Rumelhart, D., Smolensky, P., McClelland, J., and Hinton, G. 1986. Schemata and Sequential Thought Processes in PDP Models. In McClelland, J., Rumelhart, D., and the PDP Research Group, *Parallel Distributed Processing*, Vol. 2. Cambridge, MA: MIT Press.

Sellars, W. 1963. Empiricism and the Philosophy of Mind. In Sellars, *Science, Perception and Reality*. London: Routledge and Kegan Paul. Pp. 127-196.

Stich, S. 1978. Beliefs and Subdoxastic States. *Philosophy of Science*, 45, 499-518.

Wimmer, H., Hogrefe, J. and Sodian, B. 1988. A Second Stage in Children's Conception of Mental Life: Understanding Informational Access as Origins of Knowledge and Belief. In Astington, J., Harris, P. and Olson, D. (eds.), *Developing Theories of Mind*. Cambridge: Cambridge University Press. Pp. 173-192.

13

PHILOSOPHICAL ISSUES, 3
Science and Knowledge, 1993

Epistemic Folkways and Scientific Epistemology

Alvin I. Goldman

1

What is the mission of epistemology, and what is its proper methodology? Such meta–epistemological questions have been prominent in recent years, especially with the emergence ofvarious brands of "naturalistic" epistemology. In this paper, I shall reformulate and expand upon my own meta–epistemological conception (most fully articulated in Goldman, 1986), retaining many of its former ingredients while reconfiguring others. The discussion is by no means confined, though, to the meta–epistemological level. New substantive proposals will also be advanced and defended.

Let us begin, however, at the meta–epistemological level, by asking what role should be played in epistemology by our ordinary epistemic concepts and principles. By some philosophers' lights, the sole mission of epistemology is to elucidate commonsense epistemic concepts and principles: concepts like knowledge, justification, and rationality, and principles associated withthese concepts. By other philosophers' lights, this is not even part of epistemology's aim. Ordinary concepts and principles, the latter would argue, are fundamentally naive, unsystematic, and uninformed by important bodies of logic and/or mathematics. Ordinary principles and practices, for example, ignore or violate the probability calculus, which ought to be the cornerstone of epistemic rationality. Thus, on the second view, proper epistemology must neither *end* with naive principles ofjustification or rationality, nor even *begin* there.

My own stance on this issue lies somewhere between these extremes. To facilitate discussion, let us give a label to our commonsense epistemic concepts and norms; let us call them our *epistemic folkways*. In partial agreement with the first view sketched above, I would hold that *one* proper task of epistemology is to elucidate our epistemic folkways. Whatever else epistemology might proceed to do, it should at least have its roots in the concepts and practices of the folk. If these roots are utterly rejected and abandoned, by what rights would the new discipline call itself 'epistemology' at all? It may well be desirable to reform or transcend our epistemic folkways, as the second of the views sketched above recommends. But it is essential to preserve continuity; and continuity can only be recognized if we have a satisfactory characterization of ourepistemic folkways. Actually, even if one rejects the plea for continuity, a description of our epistemic folkways is in order. How would one know what to criticize, or what needs to be transcended, in the absence of such a description? So a first mission of epistemology is to describe or characterize our folkways.

Now a suitable description of these folk concepts, I believe, is likely to depend on insights from cognitive science. Indeed, identification of the semantic contours of many (if not all) concepts can profit from theoretical and empirical work in psychology and linguistics. For this reason, the task of describing or elucidating folk epistemology is a *scientific* task, at least a task that should be informed by relevant scientific research.

The second mission of epistemology, as suggested by the second view above, is the formulation of a more adequate, sound, or systematic set of epistemic norms, in some way(s) transcending our naive epistemic repertoire. How and why these folkways might be transcended, or improved upon, remains to be specified. This will partly depend on the contours of the commonsense standards that emerge from the first mission. On my view, epistemic concepts like knowledge and justification crucially invoke psychological faculties or processes. Our folk understanding, however, has a limited and tenuous grasp of the processes available to the cognitive agent. Thus, one important respect in which epistemic folkways should be transcended is by incorporating a more detailed and empirically based depiction of psychological mechanisms. Here too epistemology would seek assistance from cognitive science.

Since both missions of epistemology just delineated lean in important respects on the deliverances of science, specifically cognitive science, let us call our conception of epistemology *scientific epistemology.* Scientific epistemology, we have seen, has two branches: *descriptive* and *normative.* While descriptive scientific epistemology aims to describe our ordinary epistemic assessments, normative scientific epistemology continues the practice of making epistemic judgments, or formulating systematic principles for such judgments.[1] It is prepared to depart from our ordinary epistemic judgments, however, if and when that proves advisable. (This overall conception of epistemology closely parallels the conception of metaphysics articulated in Goldman (1989). The descriptive and normative branches of scientific epistemology are precise analogues of the descriptive and prescriptive branches of metaphysics, as conceptualized there.) In the remainder of this paper, I shall sketch and defend the particular form of descriptive scientific epistemology that I favor. (Normative scientific epistemology is discussed in the full version of the paper, of which this is an abridgement.)

[1] Normative scientific epistemology corresponds to what I elsewhere call *epistemics* (see Goldman, 1986). Although epistemics is not restricted to the assessment of *psychological* processes, that is the topic of the present paper. So we are here dealing with what I call *primary epistemics.*

2

Mainstream epistemology has concentrated much of its attention on two concepts (or terms): knowledge and justified belief. This essay focuses on the latter. We need not mark this concept exclusively by the phrase 'justified belief'. A family of phrases pick out roughly the same concept: 'well–founded belief', 'reasonable belief', 'belief based on good grounds', and so forth. I shall propose an account of this concept that is in the reliabilist tradition, but departs at a crucial juncture from other versions of reliabilism. My account has the same core idea as Ernest Sosa's *intellectual virtues* approach, but incorporates some distinctive features that improve its prospects.[2]

The basic approach is, roughly, to identify the concept ofjustified belief with the concept of belief obtained through the exercise of intellectual virtues (excellences). Beliefs acquired (or retained) through a chain of "virtuous" psychological processes qualify as justified; those acquired partly by cognitive "vices" are derogated as unjustified. This, as I say, is a *rough* account. To explain it more fully, I need to say things about the psychology of the epistemic evaluator, the possessor and deployer of the concept in question. At this stage in the development of semantical theory (which, in the future, may well be viewed as part of the "dark ages" of the subject), it is difficult to say just what the relationship is between the meaning or "content" of concepts and the form or structure oftheir mental representation. In the present case, however, I believe that an account of the form of representation can contribute to our understanding of the content, although I amunable to formulate these matters in a theoretically satisfying fashion.

The hypothesis I wish to advance is that the epistemic evaluator has a mentally stored set, or list, of cognitive virtues and vices. When asked to evaluate an actual or hypothetical case of belief, the evaluator considers the processes by which the belief was produced, and matches these against his list

[2] Sosa's approach is spelled out most fully in Sosa (1985, 1988, and 1991).

of virtues and vices. If the processes match virtues only, the belief is classified as justified. If the processes are matched partly with vices, the belief is categorized as unjustified. If a belief–forming scenario is described that features a process not on the evaluator's list of either virtues or vices, the belief may be categorized as neither justified nor unjustified, but simply *non*–justified. Alternatively (and this alternative plays an important role in mystory), the evaluator's judgment may depend on the (judged) *similarity* of the novel process to the stored virtues and vices. In other words, the "matches" in question need not be perfect.

This proposal makes two important points of contact with going theories in the psychology of concepts. First, it has some affinity to the *exemplar* approach to concept representation (cf. Medin and Schaffer, 1978; Smith and Medin, 1981; Hintzman, 1986). According to that approach, a concept is mentally represented by means of representations of its positive instances, or perhaps types of instances. For example, the representation of the concept *pants* might include a representation of a particular pair of faded blue jeans and/or a representation of the type *blue jeans*. Our approach to the concept of justification shares the spirit of this approach insofar as it posits a set of examples of virtues and vices, as opposed to a mere abstract characterization —e.g., a definition— of (intellectual) virtue or vice. A second affinity to the exemplar approach is in the appeal to a similarity, or matching, operation in the classification of new target cases. According to the exemplar approach, targets are categorized as a function of their similarity to the positive exemplars (and dissimilarity to the foils). Of course,similarity is invoked in many other approaches to concept deployment as well (see E. Smith, 1990). This makes our account of justification consonant with the psychological literature generally, whether or not it meshes specifically with the exemplar approach.

Let us now see what this hypothesis predicts for a variety of cases. To apply it, we need to make some assumptions about the lists of virtues and vices that typical evaluators mentally store. I shall assume that the virtues include belief formation based on sight, hearing, memory, reasoning in certain "approved" ways, and so forth. The vices include intellectual processes like forming beliefs by guesswork, wishful thinking, and ignoring contrary evidence. *Why* these items are placed in their respective categories remains to be explained. As indicated, I plan to explain them by reference to reliability. Since the account will therefore be, at bottom, a reliabilist type of account, it is instructive to see how it fares when applied to well-known problem cases for standard versions of reliabilism.

Consider first the demon-world case. In a certain possible world, a Cartesian demon gives people deceptive visual experiences, which systematically lead to false beliefs. Are these vision-based beliefs justified? Intuitively, they are. The demon's victims are presented with the same sorts of visual experiences that we are, and they use the same processes to produce corresponding beliefs. For most epistemic evaluators, this seems sufficient to induce the judgement that the victims' beliefs are justified. Does our account predict this result? Certainly it does. The account predicts that an epistemic evaluator will match the victims' vision-based processes to one (or more) of the items on his list of intellectual virtues, and therefore judge the victims' beliefs to be justified.

Turn next to Laurence BonJour's (1985) cases in which hypothetical agents are assumed to possess a perfectly reliable clairvoyant faculty. Although these agents form their beliefs by this reliable faculty, BonJour contends that the beliefs are not justified; and apparently most (philosophical) evaluators agree with that judgment. This result is not predicted by simple forms of reliabilism.[3] What does our present theory predict? Let us consider the four cases in two groups. In the first three cases, involving Samantha, Casper, and Maud respectively, the agent has contrary evidence that he or she ignores. Samantha has a massive amount of apparently cogent evidence that the President is in Washington,

[3] My own previous formulations of reliabilism have not been so simple. Both "What Is Justified Belief?" (Goldman, 1979) and *Epistemology and Cognition* (Goldman, 1986) had provisions —e.g., the non-undermining provision of *Epistemology and Cognition*— that could help accommodate BonJour's examples. It is not entirely clear, however, how well these qualifications succeeded with the Norman case, described below.

but she nonetheless believes (through clairvoyance) that the President is in New York City. Casper and Maud each has large amounts of ostensibly cogent evidence that he/she has noreliable clairvoyant power, but they rely on such a power nonetheless. Here our theory predicts that the evaluator will match these agent's belief–forming processes to the vice of ignoring contrary evidence. Since the processes include a vice, the beliefs will be judged to be unjustified.

BonJour's fourth case involves Norman, who has a reliable clairvoyant power but no reasons for or against the thesis tha the possesses it. When he believes, through clairvoyance, that the President is in New York City, while possessing no (other) relevant evidence, how should this belief be judged? My own assessment is less clear in this case than the other three cases. I am tempted to say that Norman's belief is *non*–justified, not that it is thoroughly *un*justified. (I construe unjustified as "having negative justificational status", and non–justified as "lacking positive justificational status".) This result is also readily predicted by our theory. On the assumption that I (and other evaluators) do not have clairvoyance on my list of virtues, the theory allows the prediction that the belief would be judged neither justified nor unjustified, merely non–justified. For those evaluators who would judge Norman's belief to be *un*justified, there is another possible explanation in terms of the theory. There is a class of putative faculties, including mental telepathy, ESP, telekinesis, and so forth that are scientifically disreputable. It is plausible that evaluators view any process of basing beliefs on the supposed deliverances of such faculties as vices. It is also plausible that these evaluators judge the process of basing one's belief on clairvoyance to be *similar* to such vices. Thus, the theory would predict that they would view a belief acquired in this way as unjustified.[4]

[4]Tom Senor presented the following example to his philosophy class at the University of Arkansas. Norman is working at his desk when out of the blue he is hit (via clairvoyance) with a very distinct and vivid impression of the President at the Empire State Building. The image is phenomenally distinct from a regular visual impression but is in some respects similar and of roughly equal force. The experience is

Finally, consider Alvin Plantinga's (1988) examples that feature disease–triggered or mind–malfunctioning processes. These include processes engendered by a brain tumor, radiation–caused processes, and the like. In each case Plantinga imagines that the process is reliable, but reports that we would not judge it to be justification conferring. My diagnosis follows the track outlined in the Norman case. At a minimum, the processes imagined by Plantinga fail to match any virtue on a typical evaluator's list. So the beliefs are at least non–justified. Furthermore, evaluators may have a prior representation of pathological processes as examples of cognitive vices. Plantinga's cases might be judged (relevantly) similar to these vices, so that the beliefs they produce would be declared unjustified.

In some of Plantinga's cases, it is further supposed that the hypothetical agent possesses countervailing evidence against his belief, which he steadfastly ignores. As noted earlier, this added element would strengthen a judgment of unjustifiedness according to our theory, because ignoring contrary evidence is an intellectual vice. Once again, then, our theory's predictions conform with reported judgments.

Let us now turn to the question of how epistemic evaluators acquire their lists of virtues and vices. What is the basis for their classification? As already indicated, my answer invokes the notion of reliability. Belief–forming processes based on vision, hearing, memory, and ("good") reasoning are deemed virtuous because they (are deemed to) produce a high ratio of true beliefs. Processes like guessing, wishful thinking, and ignoring contrary evidence are deemed vicious because they (are deemed to) produce a low ratio of true beliefs.

We need not assume that each epistemic evaluator chooses his/her catalogue of virtues and vices by direct application of the reliability test. Epistemic evaluators may partly inherit

---

so overwhelming that Norman just can't help but form the belief that the President is in New York. About half of Senor's class judged that in this case Norman justifiably believes that the President is in New York. Senor points out, in commenting on this paper, that their judgments are readily explained by the present account, because the description of the clairvoyance process makes it sufficiently similar to vision to be easily "matched" to that virtue.

their lists of virtues and vices from other speakers in the linguistic community. Nonetheless, the hypothesis is that the selection of virtues and vices rests, ultimately, on assessments of reliability.

It is not assumed, of course, that all speakers have the same lists of intellectual virtues and vices. They may have different opinions about the reliability of processes, and therefore differ in their respective lists.[5] Or they may belong to different sub–cultures in the linguistic community, which may differentially influence their lists. Philosophers sometimes seem to assume great uniformity in epistemic judgments. This assumption may stem from the fact that it is mostly the judgments of philosophers themselves that have been reported, and they are members of a fairly homogeneous sub–culture. A wider pool of "subjects" might reveal a much lower degree of uniformity. That would conform to the present theory, however, which permits individual differences in catalogues of virtues and vices, and hence in judgments of justifiedness.

If virtues and vices are selected on the basis of reliability and unreliability, respectively, why doesn't a hypothetical case introducing a novel reliable process induce an evaluator to add that process to his list of virtues, and declare the resulting belief justified? Why, for example, doesn't he add clairvoyance to his list of virtues, and rule Norman's beliefs to be justified?

I venture the following explanation. First, people seem to have a trait of *categorial conservatism*. They display a preference for "entrenched" categories, in Goodman's (1955) phraseology, and do not lightly supplement or revise their categorial schemes. An isolated single case is not enough. More specifically, merely imaginary cases do not exert much influence on categorial structures. People's cognitive systems

[5] Since some of these opinions may be true and others false, people's lists of virtues and vices may have varying degrees of accuracy. The "real" status of a trait as a virtue or vice is independent of people's opinions about that trait. However, since the enterprise of descriptive epistemology is to describeand explain evaluators' judgments, we need to advert to the traits they *believe* to be virtues or vices, i.e., the ones ontheir mental lists.

are responsive to live cases, not purely fictional ones. Philosophers encounter this when their students or non–philosophers are unimpressed with science fiction–style counterexamples. Philosophers become impatient with this response because they presume that possible cases are on a par (for counterexample purposes) with actual ones. This phenomenon testifies, however, to a psychological propensity to take an invidious attitude toward purely imaginary cases.

To the philosopher, it seems both natural and inevitable to take hypothetical cases seriously, and if necessary to restrict one's conclusions about them to specified "possible worlds". Thus, the philosopher might be inclined to hold: "If reliability is the standard of intellectual virtue, shouldn't we say that clairvoyance is a virtue *in the possible worlds* of BonJour's examples, if not a virtue in general?" This is a natural thing for philosophers to say, given their schooling, but there is no evidence that this is how people naturally think about the matter. There is no evidence that "the folk" are inclined to relativize virtues and vices to this or that possible world.

I suspect that concerted investigation (not undertaken here) would uncover ample evidence of conservatism, specifically in the normative realm. In many traditional cultures, for example, loyalty to family and friends is treated as a cardinal virtue.[6] This view of loyalty tends to persist even through changes in social and organizational climate, which undermine the value of unqualified loyalty. Members of such cultures, I suspect, would continue to view personal loyalty as a virtue even in *hypothetical* cases where the trait has stipulated unfortunate consequences.

In a slightly different vein, it is common for both critics and advocates of reliabilism to call attention to the relativity of reliability to the domain or circumstances in which the process is used. The question is therefore raised: What is the relevant domain for judging the reliability of a process? A critic like John Pollock (1986, pp. 118–119), for example, observes that color vision is reliable on earth but unreliable in the universe at large. In determining the reliability of color

[6] Thanks to Holly Smith for this example. She cites Riding (1989, chap. 6) for relevant discussion.

vision, he asks, which domain should be invoked? Finding no satisfactory reply to this question, Pollock takes this as a serious difficulty for reliabilism. Similarly, Sosa (1988 and 1991) notes that an intellectual structure or disposition can be reliable with respect one field of propositions but unreliable with respect to another; and reliable in one environment but unreliable in another. He does not view this as a difficulty for reliabilism, but concludes that any talk of intellectual virtue must be relativized to field and environment.

Neither of these conclusions seems apt, however, for purposes of *description* of our epistemic folkways. It would be a mistake to suppose that ordinary epistemic evaluators are sensitive to these issues. It is likely —or at least plausible— that our ordinary apprehension of the intellectual virtues is rough, unsystematic, and insensitive to any theoretical desirability of relativization to domain or environment. Thus, as long as we are engaged in the description of our epistemic folkways, it is no criticism of the account that it fails to explain what domain or environment is to be used. Nor is it appropriate for the account to introduce relativization where there is no evidence of relativization on the part of the folk.

Of course, we do need an explanatory story of how the folk arrive at their selected virtues and vices. And this presumably requires some reference to the domain in which reliability is judged. However, there may not be much more to the story than the fact that people determine reliability scores from the cases they personally "observe". Alternatively, they *may* regard the observed cases as a sample from which they infer a truth ratio in some wider class of cases. It is doubtful, however, that they have any precise conception of the wider class. They probably don't address this theoretical issue, and don't do (or think) anything that commits them to any particular resolution of it. It would therefore be wrong to expect descriptive epistemology to be fully specific on this dimension.

A similar point holds for the question of process individuation. It is quite possible that the folk do not have highly principled methods for individuating cognitive processes, for "slicing up" virtues and vices. If that is right, it is a mistake to insist that descriptive epistemology uncover such methods. It is no flaw in reliabilism, considered as descriptive epistemology, that it fails to unearth them. It may well be desirable to develop sharper individuation principles for purposes of normative epistemology; but the missions and requirements of descriptive and normative epistemology must be kept distinct.

This discussion has assumed throughout that the folk have lists of intellectual virtues and vices. What is the evidence for this? In the moral sphere ordinary language is rich in virtues terminology. By contrast, there are few common labels for intellectual virtues, and those that do exist —'perceptiveness', 'thoroughness', 'insightfulness', and so forth— are of limited value in the present context. I propose to identify the relevant intellectual virtues (at least those relevant to *justification*) with the belief forming capacities, faculties, or processes that would be accepted as answers to the question "How does X know?". In answer to this form of question, it is common to reply: "He saw it", "He heard it", "He remembers it", "He infers it from such–and–such evidence", and so forth. Thus, basing belief on seeing, hearing, memory, and (good) inference are in the collection of what the folk regard as intellectual virtues. Consider, for contrast, how anomalous it is to answer the question "How does X know?" with "By guesswork", "By wishful thinking", or "By ignoring contrary evidence". This indicates that *these* modes of belief formation —guessing, wishful thinking, ignoring contrary evidence— are standardly regarded as intellectual *vices*. They are not ways of obtaining knowledge, nor ways of obtaining justified belief.

Why appeal to "knowledge"–talk rather than "justification"–talk to identify the virtues? Because 'know' has a greater frequency of occurrence than 'justified', yet the two are closely related. Roughly, justified belief is belief acquired by means of the same sorts of capacities, faculties, or processes that yield knowledge in favorable circumstances (i.e., when the resulting belief is true and there are no Gettier complications, or no relevant alternatives).

To sum up the present theory, let me emphasize that it depicts justificational evaluation as involving two stages. The first stage features the acquisition by an evaluator of some set of intellectual virtues and vices. This is where reliability enters the picture. In the second stage, the evaluator applies his list of virtues and vices to decide the epistemic status of targeted beliefs. At this stage, there is no direct consideration of reliability.

There is an obvious analogy here to rule utilitarianism in the moral sphere. Another analogy worth mentioning is Kripke's (1980) theory of *eference-fixing*. According to Kripke, we can use one property to fix a reference to a certain entity, or type of entity; but once this reference has been fixed, that property may cease to play a role in identifying the entity across various possible worlds. For example, we can fix a reference to heat as the phenomenon that causes certain sensations in people. Once heat has been so picked out, this property is no longer needed, or relied upon, in identifying heat. A phenomenon can count as heat in another possible world where it doesn't cause those sensations in people. Similarly, I am proposing, we initially use reliability as a test for intellectual quality (virtue or vice status). Once the quality of a faculty or process has been determined, however, it tends to retain that status in our thinking. At any rate, it isn't reassessed each time we consider a fresh case, especially a purely imaginary and bizarre case like the demon world. Nor is quality relativized to each possible world or environment.

The present version of the virtues theory appears to be a successful variant of reliabilism, capable of accounting for most, if not all, of the most prominent counterexamples to earlier variants of reliabilism. The present approach also makes an innovation in naturalistic epistemology. Whereas earlier naturalistic epistemologists have focused exclusively on the psychology of the epistemic agent, the present paper also highlights the psychology of the epistemic evaluator.

3 References

Alston, W. (1988). "An internalist externalism". *Synthèse*, 74, 265-283.

Biederman, I. (1990). "Higher–level vision". In: D. Osherson, S.M. Kosslyn and J.M. Hollerbach, eds., *Visusal cognition and action: an invitation to cognitive science*. Cambridge, MA: MIT Press.

Biederman, I. (1987). "Recognititon–by–components: a theory of human image understanding". *Psychological Review*, 94, 115-147.

Berwick, R. (1986). "Learning from posistive only examples: the subset principle and three case studies". In: R.S. Michalski, J.G. Carbonell and T.M. Mitchell, eds., *Machine learning: an artificial intelligence approach*, vol. 2. Los Altos, CA: Morgan Kaufman.

BonJour, L. (1985). *The structure of empirical knowledge*. Cambridge, MA: Harvard University Press.

Code, L. (1987). *Epistemic responsibility*. Hanover: University Press of New England.

Fodor, J. (1983). *The modularity of mind*. Cambridge, MA: MIT Press (Bradford).

Gold, E.M. (1967). "Language identification in the limit". *Information and Control*, 10, 447-474.

Goldman, A. (forthcoming). "Review of S. Stich, *The fragmentation of reason*". *Philosophy and Phenomenological Research*.

Goodman, N. (1955). *Fact, fiction and forecast*. Cambridge, MA: Harvard University Press.

Hintzman, D. (1986). "'Schema abstraction' in a multiple–trace memory model". *Psychological Review*, 93, 411-428.

Kornblith, H. (1983). "Justified belief and epistemically responsible action". *Philosophical Review*, 92, 33-48.

Kripke, S. (1980). *Naming and necessity*. Cambridge, MA: Harvard University Press.

Medin, D.L. and M.M. Schaffer (1978). "A context theory of classification learning". *Psychological Review*, 85, 2?7-238.

Nisbett, R. and Ross, L. (1980). *Human inference: strategies and shortcomings of social judgment*. Englewood Cliffs, NJ: Prentice–Hall.

Osherson, D., Stob, M. and Weinstein, S. (1985). *Systems that learn*. Cambridge, MA: MIT Press (Bradford).

Pinker, S. (1990). "Language acquisition". In: D.N. Osherson and H. Lasnik, eds., *Language: an invitation to cognitive science.* Cambridge, MA: MIT Press (Bradford).

Plantinga, A. (1988). "Positive epistemic status and proper function". In: J.E. Tomberlin, ed., *Philosophical Perspectives*, vol. 2. Atascadero, CA: Ridgeview.

Pollock, J. (1986). *Contemporary theories of knowledge.* Totowa, NJ: Rowman & Littlefield.

Potter, M. (1990). "Remembering". In: D.N. Osherson and E.E. Smith, eds., *Thinking: an invitation to cognitive science.* Cambridge, MA: MIT Press (Bradford).

Riding, A. (1989). *Distant neighbors: a portrait of the mexicans.* New York: Vintage Books.

Smith, E.E. (1990). "Categorization". In: D.N. Osherson and E.E. Smith, eds., *Thinking: an invitation to cognitive science.* Cambridge, MA: MIT Press (Bradford).

Smith, E.E. and Medin, M. (1981). *Categories and concepts.* Cambridge, MA: Harvard University Press.

Smith, H. (1983). "Culpable ignorance". *Philosophical Review*, 92, 543-572.

Sosa, E. (1985). "Knowledge and intellectual virtue". *The Monist*, 68, 226-263.

Sosa, E. (1988). "Beyond scepticism, to the best of our knowledge". *Mind*, 97, 153-188.

Sosa, E. (1991). "Reliabilism and intellectual virtue". In: *Knowledge in perspective.* Cambridge: Cambridge University Press.

Stich, S. (1990). *The fragmentation of reason.* Cambridge, MA: MIT Press (Bradford).

Talbott, W. (1990). *The reliability of the cognitive mechanism: a mechanist account of empirical justiication.* New York: Garland Publishing.

Tversky, A. and Kahneman, D. (1973). "Availability: a heuristic for judging frequency and probability". *Cognitive Psychology*, 5, 207-232.

Tversky, A. and Kahneman, D. (1983). "Extensional versus intuitive reasoning: the conjunction fallacy in probability judgment". *Psychological Review*, 90, 293-315.

PHILOSOPHICAL ISSUES, 3
Science and Knowledge, 1993

In Defense of a Normative Scientific Epistemology

Josep E. Corbí

Alvin I. Goldman insists that epistemology[1] should be able to combine scientificity with normativity. At first glance, this proposal may sound problematic because those philosophers, like W.V. Quine and D. Campbell,[2] who stress the scientificity of epistemology also tend to envisage this discipline as a mere descriptive enterprise, with no direct bearing on normative issues. I shall argue, however, that the idea of a normative scientific epistemology not only makes sense, but defines the only available strategy. This proposal will emerge precisely from a revision of Goldman's epistemological project.

Accordingly, I shall begin by providing, in Section 1, a brief characterization of Goldman's conception of the mission and

---

[1] Although most claims defended in this paper apply to all areas in epistemology, the majority of arguments and remarks will tend to focus on what Goldman calls 'primary epistemology': "It is fully appropriate, then, for epistemology to inquire into cognitive architecture, to assess its strengths and weaknesses. At a minimum, this is a job for primary epistemology" (Goldman (1986), p. 5).

[2] Cf. Quine (1969), p. 82, and Campbell (1974), p. 413.

proper methodology of epistemology that I shall condense into three claims: the continuity thesis, the normative strain, and the scientific constraint. By contrast, in Section 2, I shall introduce a number of elements which will be used, in Section 3, to revise, ground, and qualify various aspects of Goldman's fundamental claims. These elements rest on the assumption that epistemic practices are rule-governed, and include the knowing how-knowing that distinction, a version of the holistic constraint, and what I shall call 'the principle of simultaneity' and 'the similarity assumption'. The process of evaluation and revision of Goldman's epistemological project will gradually give rise to a specific conception of normative scientific epistemology, whose main features will be summarized in Section 4.

1 Goldman's Conception of The Mission and Proper Methodology of Epistemology

In this section, I shall restrict myself to setting out Goldman's conception of scientific epistemology, which, in my opinion, can be summarized in three basic claims, namely, the continuity thesis, the normative strain, and the scientific constraint.

1.1 THE CONTINUITY THESIS

The continuity thesis claims that scientific epistemology must be 'continuous with' our ordinary epistemic concepts. I should stress, however, that, in spite of the important role that the continuity thesis plays in Goldman's epistemological project, he devotes little time to its analysis and defense. His texts do not go beyond saying that scientific epistemology should 'be continuous with' or 'have its roots in' our ordinary epistemic concepts. Such remarks simply convey the very general idea that, despite the openness to a vast range of variations, refinements, and changes, a certain degree of commonality between scientific epistemology and ordinary epistemic concepts must be guaranteed:

> To facilitate discussion, let us give a label to our commonsense epistemic concepts and norms; let us call them our *epistemic folkways*. In partial agreement with the first view sketched above, I would hold that *one* proper task of epistemology is to elucidate our epistemic folkways. Whatever else epistemology might proceed to do, it should at least have its roots in the concepts and practices of the folk. If these roots are utterly rejected and abandoned, by what rights would the new discipline call itself 'epistemology' at all? It may well be desirable to reform or transcend our epistemic folkways, as the second of the views sketched above recommends. But it is essential to preserve continuity; and continuity can only be recognized if we have a satisfactory characterization of our epistemic folkways.[3]

Moreover, Goldman hardly furnishes us with an argument to ground his continuity thesis. The quote above constitutes, in my view, the text where Goldman most clearly develops an argument to back his position. I venture his argument in this text is structured as follows: (1) Our ordinary epistemic concepts display the criteria for individuating what counts as an epistemic capacity, state, etc.(2) If scientific epistemology does not preserve at least those individuating criteria, why should we still speak of 'epistemology', except by an arbitrary nominalistic decision? (3) Hence, scientific

[3]Goldman (in press), p. 2. I should say that Goldman's actual practice as an epistemologist tends to confirm this generic understanding of the continuity thesis. Thus, we can see that an important portion of his epistemological research focuses on the analysis of our ordinary epistemic concepts. For instance, Goldman produces accurate analyses of the rules that govern our ordinary concepts of knowledge (Goldman (1986), ch. 3) and of justified belief (Goldman (1979), Goldman (1986), ch. 4 and 5). In fact, he vindicates historical reliabilism as an adaquate analysis of our ordinary concept of justified belief (Goldman (1986), p. 109 and 59). Goldman insists, however, that scientific epistemology does not reduce to the analysis of ordinary epistemic concepts; that, once our ordinary epistemic concepts have been elucidated, we should proceed to assess them and propose the corresponding revisions, refinements, and reforms (Goldman (1986), p. 1982). Thus, he seeks to show how, for instance, advances in cognitive science may induce epistemologists to refine our ordinary concept of belief (Goldman (1986), ch. 10).

epistomology must be continuous in this respect with folk epistomology, or disappearing as a discipline.

I do think, however, that Goldman's continuity thesis requires both further clarification and grounding. Accordingly, I shall seek to provide them in section 3.

1.2 THE NORMATIVE STRAIN

Goldman contends that scientific epistemology should maintain the normative strain that has traditionally been present in epistemology. The problem lies, however, in how scientificity and normativity can amalgamate in a single discipline. The continuity thesis is precisely what allows Goldman to combine the normative strain with the scientific constraint. For, according to him, the criteria to assess epistemic practices and principles derive their validity and autonomy with respect to science from the fact that they form a part of the features of our ordinary epistemic concepts whose continuity should be guaranteed in scientific epistemology, and whose validity is not affected by the development of science.[4]

From his analysis of our ordinary concept of justified belief, Goldman derives two normative expedients, namely: the reliabilist criterion of rightnesss and the reflective equilibrium method.[5] Goldman views the reliabilist criterion as the outcome of the application of the latter method, and understands that a belief-forming process is reliable if it produces a high ratio of true-beliefs, and unreliable otherwise.

---

[4] To be accurate, I should say that Goldman only explicitly endorses this line of reasoning with respect to the case of justified belief, which is the topic that he analyzes in more detail. But I assume that, 'mutatis mutandis', it could be applied to other aspects of epistemic practices.

[5] The reflective equilibrium method was initially introduced by N. Goodman (1954, pp. 20-21) and J. Rawls (1971, 46-51). To see Goldman's description of this method, cf. Goldman (1986, p. 66 and 81).

Although this method has been challenged by several objections, I consider that, on the basis of the elements to be introduced in section 2, a version of the reflective equilibrium method could easily be developed that would be able escape the most important objections addressed against it. Unfortunately, we must leave such an analysis for another occasion.

Once this notion of reliability has been introduced, we can see how, in the case of justified belief, epistemology can be both scientific and normative. Thus, given that the normative criterion of justifiedness proposed by Goldman refers, in the end, to the idea of reliability of belief-forming processes; we can, then, understand that psychology should contribute to epistemology by determining our basic belief-forming processes. At a second and independent stage,[6] the normative branch of epistemology will assess the reliability of the belief- forming processes scientifically identified and propose possible reforms. I shall challenge, in section 3, the idea that the descriptive and the normative branch of scientific epistemology should constitute two independent stages, and seek to show how scientific epistemology cannot be but normative.

1.3 THE SCIENTIFIC CONSTRAINT

Scientific epistemology is, according to Goldman, a scientifically informed enterprise in, at least, two important respects: (a) the description of our ordinary epistemic concepts and (b) the determination of our basic belief- forming processes, whose reliability scientific epistemology is supposed to assess. I should emphasize, however, that point (a) only applies to Goldman's latest view,[7] since it is explicitly denied in Goldman's previous texts.[8]

2 Elements for a revision of Goldman's stance

In this section I intend to put forward a number of elements and distinctions, which will be used in section 4 to chal-

[6] "It may well be desirable to develop sharper individuation principles for purposes of normative epistemology (a matter we shall address in Section III). But the missions and requirements of descriptive and normative epistemology must be kept distinct". (Goldman (press), 15).

[7] Cf. Goldman (in press), p. 3.

[8] Cf. Goldman (1985, p. 39) and Goldman (1986, p. 66).

lenge and revise Goldman's stance, so that a new version of the mission and proper methodology of a normative scientific epistemology should emerge.[9]

I would assume, following Goldman,[10] that epistemic practices are rule-governed. I understand by rule-governed practices those that are representational and normative. They are representational insofar as the rules that govern such practices possess a certain content, and normative inasmuch as these practices are open to being assessed as correct or incorrect by referring to the content of the rules they are governed by.

This section will be devoted to elucidate a portion of the most general principles and constraints that are involved in the individuation and assessment of rule-governed practices. This elucidation will certainly be guided by the idea of proceeding, in section 3, to a revision of Goldman's conception of a scientific epistemology.

2.1 THE KNOWING HOW-KNOWING THAT DISTINCTION

G. Ryle's distinction between *knowing how* and *knowing that*[11] focuses on a crucial feature of rule–governed prac-

---

[9] I should say that my elaboration of this point owes, as it will be obvious, a great deal to philosophers like L. Wittgenstein (1953), G. Ryle (1949), H.G. Gadamer (1975), D. Davidson (1980), Winch (1958, 1964), and D. Dennett (1987). I do not intend, though, to strictly reproduce the line of argument of any of these particular philosophers, but rather outline, out of various materials, a fairly autonomous line of reasoning that, though necessarily sketchy, should be assessed by its own merits.

[10] Cf. Goldman (1986), chapter 8, and, particularly, p. 162-3. I should say that this assumption would only be challenged by an eliminativist position (Churchland (1989), but not by naturalist approaches like Millikan's (1984), Fodor's (1987), and Dretske's (1988). In any event, in Corbí (manuscript), I discuss this issue and call into question the possibility of naturalizing intentionality.

[11] 11. Cf. Ryle (1949), ch. II. I consider that Goldman would be happy to accept some version of this distinction, since, in my view, a specific variant of it is present in the reflective equilibrium method that he assumes (Goldman (1986), p. 66).

tices that can be used, as we shall see, to qualify and defend Goldman's continuity thesis.

With respect to a speaker's linguistic abilities, we distinguish between the speaker's *knowing how* to speak and the speaker's *knowing that* such-and-such are the rules that she follows in her talk. It seems clear that mastering a language involves knowing how to speak, but not a theoretical knowledge of the grammatical rules of the language. In these circumstances, it is not surprising that competent speakers make mistakes when reflecting about their own linguistic practice, since competent speakers are not necessarily good linguists. A crucial consequence of this fact is that what counts as evidence to validate a linguistic theory is the competent speakers' intuitions about what they would say (or what they actually say) in certain contexts, but not their own intuitions about whether their behavior is generated by such–and–such a rule or such–and–such a disposition.[12]

[12]Objections could be raised, however, that we are not dealing with a neat distinction, since, for instance, knowing how to speak necessarily involves certain metalinguistic abilities. In other words, it could be argued that knowing how to use 'dog' and 'cat' includes the metalinguistic capacity for telling that 'dog' does not mean the same as 'cat', or that 'cat' refers to a sort of animal, etc. (cf. Putnam (1975)). But these last examples constitute clear cases of knowing that and, therefore, we can conclude that knowing how and knowing that are not as independent as they appeared to be.

I would not deny this contention, but I do not think it actually affects the core of my distinction. Thus, we must recall that there is all the difference in the world between the metalinguistic statements aforementioned and much more general claims like 'The meaning of a word is the idea in the speaker's head with which it is associated' or 'It is possible to create private languages', etc. All this suggests the need to distinguish different degrees of generality in a speaker's knowing that, and that, in order to acknowledge an individual as a competent speaker, a very low degree of metalinguistic knowing that is required. Moreover, the fact that a distinction presents borderline cases does not invalidate by itself the distinction in question insofar as there is an open list of cases that can be neatly placed in one of the two horns of the distinction (cf. Grice and strawson (1956), and Searle (1969, sec. 1.2.). Accordingly, I am prepared to include under the label 'knowing how' this elemental metalinguistic knowing that, and reserve 'knowing that' for the most general and principled metalinguistic knowledge. Although this distinction does

It is easy to see how our distinction between *knowing how* and *knowing that* expands beyond the case of language and encompasses all rule–governed practices. For instance, a person may be an excellent typist but not be very good at describing the rules that she follows in her typing; or she may be a good car driver and not be able to spell out the rules that she follows in her driving to orientate a beginner. Similarly, a layman may impeccably ascribe knowledge and justified beliefs to other persons, but not be very brilliant at characterizing the principles that guide his ascriptions. In section 3.1, I intend to examine in more detail how the knowing how-knowing that distinction applies to epistemic practices. Let us now shift to the holistic constraint and its corollaries.

2.2 THE HOLISTIC CONSTRAINT, THE PRINCIPLE OF SIMULTANEITY AND THE SIMILARITY ASSUMPTION

In the last decades, several versions of the presence of a holistic constraint in the individuation of content have been offered and, as one might expect, it has suffered multiple attacks and objections, which, for the sake of brevity, I cannot contemplate in this paper. Accordingly, I shall restrict myself to (i) outlining the version of the holistic constraint that I assume (but do not ground) in this paper and that, in my view, could handle the main objections raised against it;[13] and (ii) deriving two corollaries, namely, the principle of simultaneity and the similarity assumption, that will play a

not strictly conforms to Ryle's original idea, it retains part of its flavor, given that 'knowing how' is still confined to the kind of knowledge that people possess when they are good speakers and 'knowing that' refers to the elucidation of the principles involved in the linguistic practice beyond what is required to be recognised as a good practitioner.

[13]The holistic constraint was explicitly introduced by W.V. Quine (1951), it is also present in L. Wittgenstein (1953). More recently, it has been defended by D. Davidson (1980, 1984) and D. Dennett (1981, 1987). Goldman has a reluctant attitude towards certain versions of the holistic constraint and the principle of charity (Goldman (1986), sec. 8.4). I think, however, my conception of these principles is not inspired by the aprioristic perspective that he denounces and, as we shall see, is not affected by his objections to such an aprioristic approach.

significant role in our revision of Goldman's epistemological project.

The holistic constraint I subscribe to encompasses some aspects of what is normally included under the principle of charity. I understand this is not a mere terminological option, since, in my opinion, the aspects in question are usually assumed, though not stated, in most defenses of the holistic constraint. Thus, I divide the holistic constraint into two points. The first point is quite general, and says that one cannot individuate the content a subject's belief B, or a subject's utterance *U*, but against an extensive background of other beliefs, desires, skills, utterances, etc. For instance, I cannot ascribe to a subject *S* the belief 'This is a vertex of a triangle' unless she has the concept of vertex and the concept of triangle. But possessing such concepts tends to involve a certain ability to identify vertices and triangles, to perceive lines in different directions, to recognize angles, and so forth. The second point has to do with the way we assume the different elements that constitute a belief system should be articulated, and is related to an important aspect of the principle of charity. Specifically, it includes the claim that an interpreter can only individuate the content of a system of belief by assuming that the beliefs that constitute the system are basically coherent and true. To qualify a bit, I should say

---

Unfortunately, I do not have room in this paper to give a full account the main objections raised against the holistic constraint. I should simply say that I do not find convincing Fodor (1987)'s objections against the holistic constraint, which can be summarized as follows: The holistic constraint (a) goes against our commonsense intuitions about the ascription of propositional attitudes, which are atomistic; (b) rules out the possibility of ascribing to two distinct people the same belief if they do not share all their beliefs; and (c) it conflicts with the best available hypothesis to account for our cognitive capacities, namely, the Language of Thought Hypothesis.

Briefly, my reply would go as follows. With respect to (a), I should say that our commonsense knowing that about attitude ascriptions may be atomistic, but not our knowing how. With respect to (b), I consider that references to the context of comunication will allows to select the degree of background commonality required in order to ascribe the same belief to two distinct people. And, finally, I think that the emergence of the connectionist models undermines point (c) (Cf. Corbí (forthcoming)).

that I am not ruling out that there could be contradictory and/or false beliefs in the periphery of a subject's belief system, but only that a successful process of content ascription could legitimately conclude that a subject's central beliefs are inconsistent and/or false. For the latter case should be better described as a global failure in our attempt to ascribe a belief system to this particular subject.[14]

The validity of this holistic constraint should certainly be assessed by examining whether it constitutes a good or a bad reconstruction of the rules that govern our knowing how about content ascription. But, given that we have no room to develop this point now, let us take it for granted and consider two crucial corollaries that can be derived from my holistic constraint, namely: the principle of simultaneity and the similarity assumption. The principle of simultaneity maintains that the ascription of content to a subject's beliefs goes hand in hand with their evaluation. The similarity assumption claims, by contrast, that an interpreter can only individuate another subject's belief system by assuming, at the outset, that their idea of rationality, that is, their notions of coherence and truth, have important aspects in common.

The principle of simultaneity derives from the second aspect of the holistic constraint. Thus, if a necessary condition for an interpretation to be successful is that the subject's belief system does not appear as basically incoherent and false, then we cannot develop the process of interpretation, of content ascription, without at the very same time assessing whether each new ascription of content allows the system to be basically coherent and true. For, in the case that a minimal rationality were not guaranteed, we should pro-

---

[14] Now we can have a hint of why I include this second point within the holistic constraint, instead of presenting it as an aspect of the principle of charity. My idea is that the first point is too abstract, since it admits all sorts of connections between between beliefs, desires, skills, etc.. And, in this respect, it could also apply to physical and functional states, even if they hold mere causal connections. To specify the kind of connections that are proper of contents is why the second point of our holistic constraint comes into the picture. To a certain extent, our second point only seeks to render explicity what is implicit in the appeal to contents that appears in the first point of my holistic constraint.

duce the necessary readjustments in our previous ascriptions of content. These continuous movements from the parts to the whole (and inversely) and from content ascription to assessment (and inversely) have no absolute end except what the pragmatic interests that guide the entire process should advise.

To perceive the strength of the similarity assumption, we are better placed in a context of radical interpretation, where the interpreter does not know either the content of the subject's psychological states or the meaning of the subject's utterances. At this juncture, if the interpreter wants to proceed in accordance with the holistic constraint and the principle of simultaneity, she will be forced to assume that, despite eventual differences to be detected in the future, she and the subject share the central elements of their respective belief systems. Let us see why this is so.

Suppose the interpreter observes that the subject utters *U* under some circumstances, how should she interpret this utterance?. According to the holistic constraint, she should interpret this piece of behavior in such a way that it fits nicely with the rest of the subject's beliefs, utterances, etc. But how could this be done, given that, at this stage of the interpretation, nothing is known about the whole belief system. The only alternative available to start off the process of interpretation is, then, that the interpreter herself projects her own belief system onto the subject's one; in other words, that the interpreter assumes that, unless contrary evidence is available, hers and the subject's belief system have their central elements in common and, specially, their criteria of coherence and truth. This similarity assumption is not exclusive to the initial steps in the process of interpretation. It will always be present, because there will always be aspects of the subject's belief system that the interperter does not know and whose content need to be provisionally fixed in order to keep advancing in the interpretative process.[15]

---

[15] As Stich (1990), sec. 2.4, points out the similarity assumption does not set any empirical limit to how diverse belief systems can be from one another. For the similarity relation is a Sorites' relation and, therefore, belief systems in the extremes of a similarity chain could be as diverse

The similarity assumption does not exclude the possibility that, as a result of the interpretation, one could acknowledge an important degree of disparity between the interpreter's and the subject's principles of rationality. Similarly, the holistic constraint and the similarity assumption do not rule out either the possibility of individuating irrational beliefs, inferences, or practices. When, on the basis of a given interpretation, we consent to ascribe a number of irrationalities to a subject; then we may face two quite different cases: either the irrationalities in question do not affect the kernel of the belief system or they do. In the former case, there is no conflict with the holistic constraint and its corollaries. This is the case, for instance, of a practitioner's mistakes in her theoretical intuitions about her knowing how in any epistemic practice. If, by contrast, the irrationalities do affect the kernel of the belief system, then our holistic constraint surely forces us to give up viewing this subject as holding any belief system at all. Think of madness, for example.[16]

2.3 INTENTIONAL VS. FUNCTIONAL THEORIES

Along the lines of the preceding subsection, I could say that intentional theories are those that seek to describe, explain

---

as you wish. Nevertheless, the similarity assumption does set a limit to how diverse belief systems can be if they are to be individuated by a given interpreter at a certain time, and this is all I need in this paper. Of course, this similarity assumption implies that there can be belief systems that a given intepreter cannot individuate.

[16] This is the way I would suggest we should deal with the irrational inferences pointed out by Nisbett and Ross (1980), Kahneman, Slovic, and Twersky (1982), and others. For a bit of philosophical discussion on this issue, cf. Cohen (1986, 1991) and Stich (1990, 1991a, 1991b).

I do not think Goldman (1986, p. 176)'s case of massive falseness affects may position. For I am not interested in the quantity, but in the quality, of the beliefs that could prove to be false. I am not saying that the majority of a subject's beliefs cannot be false, but that her basic beliefs cannot be false. In my view, Goldman's case of massive confusion is intelligible because we share the subject's basic beliefs about how to know things about the world and the subject is wrong only about quite peripheral beliefs, namely, those that one can learn in newspapers and encyclopedias.

and predict rule-governed practices. The object-matter of intentional theories conditions the nature of their empirical research, since the recollection of data in this kind of theory must be guided by the holistic constraint, the simultaneity principle, and the similarity assumption. In other words, my contention is that, despite appearances to the contrary, these three points are central to sciences like economics, sociology, psychology, etc., which are devoted to the study of rule-governed practices.

By contrast, functional theories move at a different level of description. If intentional theories are basically associated with the idea of rule, functional theories are concerned with the notion of mechanism that admits to multiple physical instantiations. Given this latter fact, it is clear that functional properties do not reduce to physical properties, since they are to be individuated differently: the functional properties of a state, event or object depend on the causal role that a functional theory ascribes to it. Accordingly, we can say that the object-matter of functional theories is to explain and predict the outputs of a mechanism, independently of their physical properties; the states and objects of the mechanism in question being individuated by the causal role that the functional theory itself adscribes to them.

3 Revisiting Goldman's stance

Now the time is ripe to revisit the main theses that define Goldman's stance concerning the idea of a normative scientific epistemology, and see whether they can be maintained or they should, by contrast, be reformed, refined, or rejected.

3.1 THE CONTINUITY THESIS REVISITED

I intend to revise Goldman's continuity thesis and defend a qualified version of it on the basis of the elements furnished in the previous section. Thus, resuming the knowing how-knowing that distinction, I propose the label 'folk epistemic practice' to refer to commonsense epistemic knowing how, and reserve 'folk epistemic theoretical intuitions' for commonsense intuitions about quite general and theoretical questions concerning epistemic concepts. Accordingly, the issue about the continuity thesis should be divided into two basic questions: (a) What is the role that folk epistemic practice should play in scientific epistemology? (b) What is the role that folk epistemic theories should play in scientific epistemology?. Facing these questions, I should say that, although I have not yet supplied a full argument, I think I have at least suggested, in section 2.1, why it may be plausible to vindicate the continuity thesis with respect to folk epistemic practices, but not with regard to folk epistemic theoretical intuitions. Accordingly, I suggest that we should reinterpret Goldman's continuity thesis and his argument to back it as being exclusively concerned with folk epistemic practice. We pointed out, however, that the first premiss of Goldman's argument was in need of further grounding, and this what I plan to provide now.

I would like to divide the issue concerning the continuity thesis between folk epistemic practice and scientific epistemology into two subquestions: (1) Why epistemic theory should be continuous with *epistemic practice* in general, and (2) Why epistemic theory should be continuous with *our folk* epistemic practice?. I think a positive response could easily be provided to (1), thanks to the distinction between knowing how and knowing that.

Of course, objections could be raised that the most that such an argument could prove is that scientific epistemology needs to be continuous with some sort of epistemic practice, namely, scientific epistemic practice. But nothing in the knowing that-knowing how distinction would allow us to guarantee that scientific epistemology should be continuous with our folk epistemic practice. I think this is true and, in fact, this objection points in the direction suggested by question (2). To emphasize the importance of this question, I should say that we have not even guaranteed the existence of a basic continuity across the folk epistemic practices that are present in the distinct cultures and historical periods; that nothing has been said either to ensure that folk epistemic practice is to be continuous with scientific, aesthetic, or mystic epistemic practices. Accordingly, what is required is an argument that assures the continuity across all sorts of epistemic practices.

I understand that the continuity across subjects and cultures can be guaranteed by the similarity assumption, which claims that an interpreter can only individuate another subject's belief system by assuming that they share their basic criteria of rationality. Hence, when a scientific epistemologist seeks to work out the principles that govern a certain epistemic practice, she is also bound to adopt the similarity assumption and, therefore, to assume a similarity, a continuity, between the criteria of rationality of her own epistemic practices and those of any other epistemic practices that she could individuate.

However, the similarity assumption does not assure the continuity across *all* the epistemic practices in which a subject is involved. Still, our holistic constraint forces us to assume, at least, a certain continuity across a subject's *central* epistemic practices, since, otherwise, we would not be able to recognise this subject's beliefs as constituting a belief system that is basically coherent and true.

3.2 THE NORMATIVE STRAIN REVISITED

I am certainly interested in maintaining a normative strain within scientific epistemology. I think, however, that Goldman's position requires (i) a bit of clarification and grounding concerning the status of his reliabilist criterion of rightness, and (ii) a revision of his sharp separation of the normative and the descriptive branches of scientific epistemology.

Goldman claims that his reliabilist criterion is the result of the elucidation of our folk epistemic practice.[17] Nevertheless, no argument is provided to justify why this normative standard is not a mere internal criterion of justifiedness, specific to our folk epistemic practice; but a criterion that could be used in the assessment of the justifiedness of the beliefs involved in other folks' epistemic practices. Unfortunately, if an argument of this sort is not provided, Gold-

[17] At this juncture, I am adapting Goldman's claim to the terminology coined in this paper.

man will be reduced, despite his claims on the contrary, to a relativistic position.[18] In the preceding subsection, I have shown, however, how the similarity assumption may help us to provide the argument that Goldman requires to preserve an interesting normative strain. For, on the basis of this assumption, we can assure that if a reliabilist criterion of rightness proved to be crucial to our central epistemic practices; then, we could conclude that this criterion will also be present in the central epistemic practices of any subject whose belief system we can individuate.[19] In such a case, Goldman would be entitled to vindicate his reliabilist criterion as a normative standard to assess not only the justifiedness of our ordinary beliefs, but the justifiedness of the beliefs implicated in the central epistemic practices of any subject's belief system.

Let us now turn to the relationship between the descriptive and normative branches of scientific epistemology. My contention is that the principle of simultaneity rules out Goldman's division of the task of a scientific epistemology into two stages: first descriptive and then normative. What the principle of simultaneity stipulates is that the description of rule-governed practices and, therefore, of epistemic practices, cannot be carried out without at the same time assessing them, that description and assessement go hand in hand in so far as rule-governed practices are concerned. This fact is certainly relevant because it shows how deeply the normative strain is rooted in scientific epistemology (and, in general, in all intentional theories). For, from this perspective, it is no longer possible to stop the epistemological enterprise at the descriptive stage, since each step that you can take in epistemology is full of normative judgements. To put it shortly, I am not simply claiming, as Goldman does, that scientific epistemology *can* preserve the traditional normative strain; but that scientific epistemology *must* be normative.

[18] Cf. Goldman (1986), p. 70.

[19] Of course, we can always retrospectively revise our judgements about whether a certain feature is or is not crucial to our epistemic practice.

3.3 THE SCIENTIFIC PRINCIPLE REVISITED

I wish to examine now Goldman's scientific principle, which could be summarized as follows: Science ought to contribute to scientific epistemology in a double way: (1) The task of elucidating or describing our ordinary epistemic concepts is itself a scientifically informed task and (2) the results of cognitive science are relevant to determine the basic belief-forming processes to which scientific epistemology should apply the normative standards detected in (1).[20]

I am entirely prepared to grant point (1), since I view the elucidation of our ordinary concepts as a task proper of an intentional theory, and whose performance is constrained by the same conditions as any empirical investigation about rule-governed practices. I consider, by contrast, that point (2) requires a bit of clarification.

The extent to which cognitive science can be relevant to scientific epistemology will depend on our understanding of the level of description at which cognitive science should be placed. More concretely, whether it operates at a functional or at an intentional level.[21] As we have seen, the idea of a mechanism designed or selected to perform a certain function will be essential to the former; whereas the notions of representation and misrepresentation, of correctness and incorrectness, are at the core of the latter.[22]

Suppose that we restrict cognitive science to the functional level, then, given that scientific epistemology operates at an intentional level, the most that cognitive science can contribute to scientific epistemology is by indicating lower-level constraints upon what can be true at an intentional level.

[20] Of course, this remark refers basically to justified beliefs.

[21] To simplify, I do not mention the neurological level, but a similar remark could be produced with respect to this lower level. In any event, Goldman himself tends also to neglect neurological theories.

[22] There are, of course, several theories about how these two levels relate and, particularly, about whether the intentional level could be reduced to the functional one. Unfortunately, I have no room in this paper to discuss this issue and, in what follows, I shall confine myself to examining the consequences for contribution of cognitive science to scientific epistemology of the discrimination between functional and intentional theories, which I introduced in Sec. 2.3.

Accordingly, cognitive science will not be able to fix the basic belief-forming processes, since the notion of belief only makes sense within an intentional approach.[23]

If, by contrast, we incorporate intentional psychology to cognitive science, then the contribution of cognitive science will be particularly important. For in such a case, cognitive science could provide us, for instance, with an intentional theory where an organism's basic belief-forming processes could be determined and their reliability assessed. This fact should not mislead us into thinking, though, that normative scientific epistemology would be reduced to intentional cognitive science. For, even if they share the characteristics that define intentional theories, their respective object-matters and emphases tend to differ. Thus, normative scientific epistemology (1) encompasses not only epistemic capacities of individuals, but the social dimension of knowledge, (2) is not interested in all sorts of psychological processes, but only in those that are epistemically relevant, and (3) tends to stress the normative dimension, that is, the evaluation of epistemic practices and the proposal of alternatives and reforms.

4 Conclusion: a normative scientific epistemology

In the light of the previous comments on Goldman's three basic claims concerning the mission and proper methodology of scientific epistemology, I think I can summarize my own conception of the basic features of a normative scientific epistemology as follows:

(i) *Normative scientific epistemology is bound to be continuous with our folk epistemic practice.* For all epistemic theories need to rely on one or another epistemic practice; and,

[23] Similarly, cognitive science will not be able to determine, at a functional level, whether a certain cognitive process is or is not reliable, since the notion of reliability only makes sense within the intentional stance. For, as we have seen, the degree of reliability of a process will depend on the ratio of true beliefs that it brings about; but the notion of true belief can only be interpreted in the context of the notions of representation and misrepresentation.

besides, the similarity assumption, together with the holistic constraint, guarantees that all central epistemic practices that we can individuated must be continuous with our own central epistemic practices.

(ii) *Normative scientific epistemology not only can but must be normative.* Scientific epistemology can be normative because the similarity assumption, together with the holistic constraint, guarantees a coincidence in the basic rules that govern the any subject's central epistemic practices. Moreover, scientific epistemology *must* be normative because the principle of simultaneity dictates that the description of rule-governed practices goes necessarily hand in hand with its evaluation.

(iii) *Normative scientific epistemology is scientifically informed in a pair of ways.* Thus, scientific epistemology must (a) take into consideration the constraints that our knowledge at a neural and a functional level impose upon what can be true at an intentional level; and (b) be viewed as a discipline that, relying on several intentional sciences, focuses on epistemic practices both form an individualistic and a social perspective, and tends to emphasize the normative evaluation and the proposal of reforms for our epistemic capacities and practices.[24]

[24]This paper has benefited from comments by Alvin I. Goldman, Nicolás Sánchez, Carlos J. Moya, Julián Marrades, and S. Stich.

5 REFERENCES

Campbell, D.: 1974, "Evolutionary Epistemology" in Schilpp, R. (ed.), *The Philosophy of Karl Popper*, La Salle, Ill.: Open Court, vol. 1.

Churchland, P.M.: 1989, *A Neurocomputational Perspective*, Cambridge, Mass.: The MIT Press.

Cohen, J: 1986, *The Dialogue of Reason*, Oxford: Oxford University Press.

Cohen, J.: 1991, "Stephen P. Stich, The Fragmentation of Reason" in *Philosophy and Phenomenological Research* 51, 185-188.

Corbi, J.E.: (manuscript), "Failing to Naturalize Intentionality".

Corbi, J.E: (forthcoming), "Classical and Connectionist Models: Levels of Description" in *Synthese*.

Davidson, D.: 1980, *Essays on Actions and Events*, Oxford: Clarendon Press.

Davidson, D.: 1984, *Inquiries into Truth and Interpretation*, Oxford: Clarendon Press.

Dennett, D.: 1981, *Brainstorms*, Brighton, Sussex: The Harvester Press.

Dennett, D.: 1987, *The Intentional Stance*, Cambridge, Mass.: The MIT Press.

Dretske, D.: 1988, *Explaining Behavior*, Cambridge, Mass.: The MIT Press.

Fodor, J.: 1987, *Psychosemantics*, Cambridge, Mass.: The MIT Press.

Gadamer, H.G.: 1975, *Wahrheit und Methode*, Tübingen: J.C.B. Mohr (Paul Siebeck).

Goldman, A.I.: 1979, "What is Justified Belief?" in G.S. Pappas (ed.), *Justification and Knowledge*, Dordrecht: D. Reidel Publishing Company, 1-23

Goldman, A.I.: 1986, *Epistemology and Cognition*, Cambridge, Mass.: Harvard University Press.

Goldman, A.I.: 1985, "The Relation Between Epistemology and Psychology", *Synthese* 64, 29-68.

Goldman, A.I.: (in press), "Epistemic Folkways and Scientific Epistemology" in Goldman, A.I., *Liaisons: Philosophy Meets The Cognitive and Social Sciences*, Cambridge, Mass.: The MIT Press.

Goodman, N.: 1954, *Fact, Fiction, and Forecast*, Cambridge, Mass.: Harvard University Press.

Grice, H.P. & Strawson, P.F.: 1956, "In Defense of a Dogma" in *Philosophical Review* 65, 141-158.

Kahneman, D.; Slovic, P.; and Tversky, A.: (1982), *Judgement and Uncertainty: Heuristics and Biases*, Cambridge: Cambridge University Press.

Millikan, R.G.: 1984, *Language, Thought, and Other Biological Categories*, Cambridge, Mass.: The MIT Press.

Nisbett, R. & Ross, L.: 1980, *Human Inference: Strategies and Shortcomings of Social Judgement*, Englewood Cliffs, N.J.: Prentice-Hall.

Putnam, H.: 1975, "The Meaning of 'Meaning'" in Putnam, H., *Philosophical Papers*, v. 2: Mind, Language, and Reality, Cambridge: Cambridge University Press.

Quine, W.V.: 1951, "Two Dogmas of Empiricism" in *Philosophical Review*, January.

Quine, W.V.: 1969, "Epistemology Naturalized" in Quine, W.V., *Ontological Relativity and Other Essays*, New York, Columbia University Press.

Rawls, J.: 1971, *A Theory of Justice*, Cambridge, Mass.: Harvard University Press.

Ryle, G.: 1949, *The Concept of Mind*, London: Hutchinson.

Searle, J.R.: 1969, *Speech Acts. An Essay in the Philosophy of Language*, Cambridge: Cambridge University Press.

Stich, S.P.: 1990, *The Fragmentation of Reason*, Cambridge, Mass.: The MIT Press.

Stich, S.P.: 1991a, "The Fragmentation of Reason: Précis of Two Chapters", *Philosophy and Phenomenological Research*, 51, 1, 179-183.

Stich, S.P.: 1991b, "Evaluating Cognitive Strategies: A Reply to Cohen, Goldman, Harman, and Lycan", *Philosophy and Phenomenological Research*, 51, 1, 207-213.

Winch, P.: 1958, *The Idea of a Social Science*, London: Routledge and Kegan Paul.

Winch, P.: 1964, "Understanding a Primitive Society" in *American Philosophical Quarterly*, I, 307-24

Wittgenstein, L.: 1953, *Philosophical Investigations*, Oxford: Basil Blackwell.

15

PHILOSOPHICAL ISSUES, 3
Science and Knowledge, 1993

Epistemic Aims and Values in W.V. Quine's Naturalized Epistemology[1]

Javier Rodríguez Alcázar

1 The problem

It is usually claimed that certain aims and values have a decisive influence on scientific activity, while it is not the business of science to determine which of these aims and values ought to guide its behaviour. I intend to raise some questions regarding this supposed *external* character of aims and values. It is not my purpose, then, to take sides in any particular debate on which values and goals should direct scientific activity. I merely hope to inspire you to reflect upon the *field* in which debates of this kind take place.

The scope of my paper will be limited to the so-called *epistemic* or *cognitive* aims and values. I shall not, therefore,

[1]I am indebted to David O. Brink, María José Frápolli, Esther Romero and participants at the 1991 SOFIA Conference for their comments on different versions of this paper.

bring into consideration those *moral* aims and values which also influence the direction of science to a great extent.[2] Simplicity and predictive power are two well-known examples within epistemic ends and values:[3] we sometimes say that *simplicity* is an epistemic value which may be relevant when choosing one theory formulation among several. And it is also usually said that one of the goals of science is the *prediction* of events.

I intend, therefore, to discuss whether the debates about cognitive aims and values are outside science or if they may be approached from within and using its methods.

sectionValues and Quine's naturalism

I shall approach this topic from the context of W. V. Quine's version of *epistemological naturalism.* This perspective is particularly interesting as certain central elements of Quine's philosophy are not coherent with one viewpoint held by many philosophers of science regarding the debates on epistemic values and aims. Strange as it may seem, though, Quine shares this very attitude himself.

These philosophers would all deny that the discussion on aims and values may take place *within* science. One conspicuous example of such a stance is found in the work of Karl Popper, who suggests that there is no rational way of facing disagreements on cognitive aims, since the choice of one scientific aim in favor of another is purely a subjective and emotive matter.[4]

---

[2] One debate ocurring in this moral sphere concerns, for instance, whether the main purpose of science is to satisfy our quest for knowledge or to provide us with technology; another, whether progress in our knowledge of the world should be limited when this progress is potentially menacing for human beings.

[3] These are instances of both aims and values. The same object may be an aim or a value, depnding on the way we look at it.

[4] An example is to be found in §9 of *The Logic of Scientific Discovery*, in which Popper begins the search for a 'criterion of demarcation' between empirical science and other enterprises. He asserts there that rational discussion is only possible when the sides id dispute share some purpose. As a consequence of this, his own proposal of a certain epistemic end as defining scientific activity has to be regarded as something outside both science and rational debate. As Popper says, his criterion of demarcation should be seen "as a *proposal for an agreement or con-*

Quine, as we shall see, does not exactly affirm that the debate on epistemic aims goes before science. Rather, he tends to ignore the existence of this kind of controversy; in other words, Quine seems to assume that those taking part in the science game agree all the way on the purposes of the game.[5] Anyway, he shares the classical viewpoint that throws values out of science. For instance, at the beginning of §13 of *Roots of Reference* Quine asserts: "Scientific theory stands proudly and notoriously aloof from value judgements".[6] This emphatic statement clearly shows that, although Quine, as we all know, has softened positivists' requirements for scientific objectivity, he still believes, as they did, that this objectivity is unattainable in the contested realm of aims and values. One immediate consequence of this attitude is the following thesis, the first of three I wish to bring into consideration:

Thesis 1. Science cannot take charge of the debate on the ends and values of science itself.

As I will try to show next, however, the basic Quinean commitment to naturalism is at risk as soon as we accept this thesis; to be precise, I will argue that the only way of making this thesis compatible with Quinean naturalism is by assuming that Quine's naturalistic epistemology is disabled as far as accomplishing normative tasks is concerned. More schematically, the above thesis clashes with the combination of two other Quinean theses which we may sum up as follows:

Thesis 2. Epistemology is a chapter of science.

Thesis 3. Epistemology is a normative enterprise.

Thesis 2 is a corollary of Quinean naturalism, which, in my opinion and that of more authoritative commentators,

---

vention", and the choice of the epistemic goal as "a matter of decision, going beyond rational argument". See Popper, K. (1997^9) *The Logic of Scientific Discovery.* London: Hutchinson, p. 37.

[5]This assumtion, by the way, is easily refuted by a glance at the history of science.

[6]Quine, W.V. (1974) *Roots of Reference.* La Salle: Open Court, p. 49.

constitutes the very foundation of what we unhesitatetingly call Quine's epistemological system.[7] Naturalism, according to Quine's precise use of the term, is the doctrine which states that no external discipline has anything to add to the description of reality provided by science. Now, human beings, together with their actions and products, are included in that reality; it follows that science itself is part of the reality described by science. Therefore, everything about science has to be said by a *science of science* (and this is precisely Quine's "naturalized epistemology"). In this way Quine signs the death warrant of the *first philosophy*, and as a consequence he dictates the transformation of the epistemological questions into problems which can be solved within the framework of science and using scientific means.[8] And this is exactly what Thesis 2 states.

The third thesis at stake emerges as a reaction against an indictment often levelled at Quine.[9] The general accusation may be summarized as follows: a *scientific* epistemology like Quine's is merely a *descriptive* discipline, and, given that the prescription of the best methodological rules for the acquisition and justification of scientific knowledge is an essential component of a real epistemology, we might conclude that Quine succeeds in naturalizing just one part of the theory of

---

[7] Perhaps the most convincing defender of the basic character of naturalism within Quine's philosophical system is Roger Gibson. See Gibson (1982) *The Philosophy of W.V. Quine: An Expository Essay*. Tampa: University of South Florida Press, and Gibson (1988) *Enlightened Empiricism: An Examination of W.V. Quine's Theory of Knowledge*. Tampa: University Presses of Florida.

[8] See "Things and Their Place in Theories", in Quine (1981) *Theories and Things*. Cambridge, MA: The Belknap Press of Harvard University Press, p. 21.

[9] One among many examples is Jaewon Kim, according to whom Quinean epistemology is purely descriptive. For him, this is shown by the fact that the notion of *justification* (which is *par excellence* the normative notion of epistemology) has gone by the board in Quine's project. Naturalized epistemology, Kim points out, can offer no more than a *description* of the *causal relation* which exists between the "meager input" of sensory stimulation and the "torrential output" consisting of a complex scientific building of beliefs. Cf. Kim, Jaewon (1988) "What is 'Naturalized Epistemology'?", in Tomberlin, J.E. (ed.), *Philosophical Perspectives*, 2. Atascadero: Ridgeview, pp. 381-405.

knowledge. Naturalized epistemology, therefore, would not be a genuine alternative to traditional epistemology.

Quine reacts to this charge by claiming that, in addition to its descriptive side, naturalized epistemology has a normative one: there is, then, no *legitimate* task among those allegedly carried out by classical theory of knowledge that the Quinean epistemologist should refrain from undertaking. Quine, of course, holds that the "new" epistemology does not attempt to fulfil certain tasks, such as building the philosophical foundations of scientific knowledge, which the naturalist philosopher judges to be spurious. But he claims that the two main purposes of the traditional theory of knowledge (i.e., the description of the theory acquisition process *and* the justification of scientific standards) are accomplished much more successfully by naturalistic epistemology. In short, what I call Thesis 3 is the assertion that naturalized epistemology is, after all, one way (the only sensible way) of doing *epistemology.*[10]

When we attempt to combine the above three theses, as Quine does, they collide with one another. The collision occurs as follows. One consequence of the combination of Theses 2 and 3 is that normative talk is a part of scientific discourse. But the insertion of normative talk into science apparently necessitates the introduction of goals and values into the framework of scientific negotiation (for instance, we deem the epistemic norm of recommending simple theory formulations as justified because we *value* simplicity highly or because we take simplicity as one of the *aims* to be pursued by our scientific methodology). And the admission of aims and values in science is incompatible with the content of Thesis 1.

---

[10]In three places, at least, Quine explicitly asserts the normative weight of his empiricist convictions: see Quine's "Comment on Lauener", in Barret and Gibson (eds.) (1990) *Perspectives on Quine*, Oxford: Blackwell, p. 229; see also pp. 39 and 181 of *Theories and Things, cit.* In the latter ("Responding to M.J. Cresswell"), Quine holds that the *scientific* finding that information reaches us only through our nerve endings has normative force, because it discredits other purported channels for knowledge acquisition.

If I am right, then, Quine ought to abandon one of the three theses above. The possible solutions are:

(i) to relinquish Thesis 1; this solution implies the broadening of our concept of science to the point of including in it a *genuinely* normative discourse, i.e., one which may embrace the debate on the objectives and values guiding scientific work,

(ii) to give up Thesis 2 and, consequently, *naturalism* itself. This would mean that normative talk about science takes place within the limits of a portion of epistemology which is external to science, or

(iii) to renounce Thesis 3 and, as a result, the claim that naturalistic epistemology may carry out *normative* tasks. This option would lead us to something similar to epistemological behaviorism à la Rorty, which means the disappearance of epistemology in the proper sense of the term.[11]

Quine, however, resorts to an argument which *apparently* allows him to avoid plumping for any one of these solutions. After describing Quine's strategy, I shall endeavour to show that he has indeed unintentionally given up one of the aforementioned theses, and that the thesis he abandons is the very one which he should take most care to keep intact in order to preserve the coherence of his philosophical system. In other words, the surrendered thesis is *naturalism*.

2 Quine's move

Quine defends himself by denying that normative recommendations inside science require bringing aims and values into the theatre of scientific debate. In turn, this denial is supported by a description of normative talk as a kind of "en-

[11] See Rorty (1979) *Philosophy and the Mirror of Nature*. Princeton: Princeton University Press, sp. §IV.2.

gineering" or technological discourse.[12] A technological discourse is, in this context, the kind of talk we use to find out the most efficient *means* of achieving a certain objective.[13] A discourse of this kind is patently normative in character, but it becomes purely descriptive as soon as *we are given* the last parameter. And this, according to Quine, is the case in normative epistemology: since the goal of scientific "language game" is already given, epistemology can fulfil its normative duties by merely *describing* those practices which do in fact best promote the desired goal. If we ask Quine what the indisputed aim of science is, Quine's answer is as categorical as it is commonplace: the epistemic end *par excellence* is the *prediction* of sensory stimulation.[14]

Quine has completed, then, his move: scientific epistemology, held aloof from value judgements, can happily give normative advice since this is one result of combining two elements which are not normative at all: the first is the empirical description of the procedures which best achieve the epistemic aims of science; the second is the identification of these cognitive aims, which Quine seems to reduce to one, i.e., the anticipation of sensory stimulation. For, in Quine's opinion, this identification does not have the character of a normative recommendation —it is merely a matter of identifying the supreme rule of a certain "language game" whose norms, as in other games, are given to and accepted by the players before they start.[15]

---

[12] Cf. Quine (1990) *Pursuit of Truth.* Cambridge, MA, and London: Harvard University Press, p. 19. See also Quine's "Reply to Morton White", in Hahn and Schlipp (eds.) (1986) *The Philosophy of W.V. Quine.* La Salle, Ill.: Open Court, pp. 664-665.

[13] Cf. Quine's "Reply to Morton White", in Hahn and Schlipp (eds.), *op. cit.*, p. 665.

[14] See pp. 2 and 20 of *Pursuit of Truth,* in which Quine distinguishes between the "purposes" of the scientific language game (which I labelled above as the "moral" aims of science) and "what decides the game" (i.e., the"cognitive" aim of science, in the terminology I have adopted). The former are, in Quine's opinion, understanding and technology. The latter is identified as the prediction of future sensory input. Cf. also *Roots of Reference,* p. 137, and "Comment on Lauener", p. 229.

[15] See *Pursuit of Truth,* p. 20.

Unlike *ethics*,[16] which is a *genuinely* normative discipline because of the debatable character of the objectives to be pursued, the aim of science is, according to Quine, given *by definition*. It follows that the epistemic aim of scientific research, unlike every scientific statement, is over and above discussion and revision.[17] That aim, then, established in a sphere outside science, is *a priori* to scientific work.

3 A criticism and some suggestions

These conclusions are, of course, unsatisfactory. In the first place, Quine's establishing of a certain cognitive aim as defining the language game of science brings to mind the attitude of a dogmatic moralist who would establish that, also by definition, the goal of moral acts is, for instance, reward in heaven. He would be quite willing to discuss the means towards that end, but not the legitimacy of alternative goals. Secondly, and more importantly, conclusions of this kind are at variance with some of the ideas Quine has instilled upon us for decades. They are at odds with his rejection of *a priori* sources of knowledge. They run counter to his skeptical attitude towards the fruitfulness of a philosophical epistemology devoted to the *analysis* of epistemic terms.[18] And, finally,

---

[16] In his "Reply to Morton White" (Hahn and Schlipp (eds.), *op. cit.*, p. 665) Quine explicitly contrasts the debatable character of the moral aims and the situation in science.

[17] Quine asserts that, even if we were forced to give up empiricism under the pressure of very strong evidence to the contrary, "the test of the resulting science would still be predicted sensation" (*Pursuit ofTruth*, p. 21).

[18] We find an example of Quine's mistrust towards the possibilities of an "analytical" epistemology in his attitude regarding the attempts to analyse the term 'knowledge'. Cf. Quine (1984) "Relativism and Absolutism", *The Monist*, 67, p. 295. Cf. also Quine (1987) *Quiddities: An Intermittently Philosophical Dictionary*, Cambridge, MA, and London: The Belknap Press of Harvard University Press, p. 108. Quine's general position on the matter is that the "analytical" epistemology should be replaced by a "naturalistic" one, that is, by the scientific study of the theory construction process. His claims on the cognitive goal of science, however, seem to be a result of analyzing the term 'science'; and, in my

these conclusions clash with Thesis 2 above. According to this thesis, epistemology is a chapter of science. But all scientific hypothesis and theories are, in principle at least, open to revision.[19] What can we say, then, about the defining of the immutable essence of scientific knowledge? This would only seem possible in the context of a "first philosophy" and does not fit, therefore, into naturalized epistemology.[20]

I wish to stress that I have no intention of discussing whether the main epistemic goal of science is to provide reliable predictions. But I do question whether a *naturalist* may accept that a belief such as this is less subject to discussion and revision than the laws of, say, thermodynamics, arithmetic or logic (for all of them are, at least *in principle*, revokable, according to Quine's holism and naturalism).[21]

Unavoidably, then, we must discard at least one of the three theses I have formulated above. I will not attempt to argue in favor of the relinquishing of any one of them. Nevertheless, from the point of view of a *naturalist*, the least satisfactory solution is exactly the one which is concealed behind the Quinean move I have described; that solution demands the surrendering of Thesis 2. On the other hand, the naturalist practitioners of epistemology who reject a concept of the theory of knowledge as merely a *descriptive* enterprise (i.e., those who reject solution (iii)) have good reason to be sympathetic towards the rejection of Thesis 1. This solution entails the admittance of the debate on the aims and values

---

opinion, Quine ought to be as skeptical towards the fruitfulness of this analysis as he is towards that of the analysis of the term 'knowledge'.

[19] For an explicit Quinean admission of the fallible and corregible character of science, see *Pursuit of Truth*, p. 21.

[20] One possible defenseon Quine's part could be to claim that his definition of the "language game" of science as an activity aimed at succesful prediction merely describes what the scientific community inderstands by 'science'. But this would place Quine in the position of an "epistemological behaviourist" who merely *describes* the current social practices and refrains from making normative criticisms to them.

[21] Cf., for instance, the well-known Quinean statements on the consequences of holism in §6 of "Two Dogmas of Empiricism", in Quine (1953) *From a Logical Point of View*, Cambridge, MA: Harvard University Press.

of science into naturalized epistemology and therefore into science itself.

If Quine were to accept this solution, he would also be able to accept that whenever he speaks of the normative force of empiricism,[22] recommends simplicity, conservatism and other epistemic virtues of hypotheses[23] or makes normative judgements on what entities may be admitted into naturalistic ontology,[24] he is using a genuinely normative and evaluative (and not merely "technological") discourse. And accepting this would save Quine, by means of a true *radicalization* of his naturalism, from becoming a secret practitioner of the "first philosophy".

I will not attempt to show how the debate on aims and values would be possible in the context of a naturalistic epistemology. But I would suggest that the naturalistic philosopher should look to the history of science for some help.[25] History shows how scientists down the centuries have been faced with the task of harmonizing their particular theories, their methodological norms and their cognitive aims and values in an endless process affecting not only the content of theories and the rules of scientific method, but also the set of accepted aims and values and, along with them, the very definition of science itself.[26]

---

[22] See n. 9 above.

[23] See Quine, W.V., and Ullian, J.S. (1978) *The Web of Belief.* New York: Random House, pp. 66 ff. See also *Roots of Reference*, pp. 137-138.

[24] As in §§34, 35 and 36 of *Roots of Reference* and in §50 of *Word and Object.* Cambridge, Mass.: MIT Press, 1960.

[25] See, for an illuminating description of these historical processes, Laudan, L. (1984) *Science and Values.* Berkeley, LA, and London: University of California Press.

[26] I want to thank to Mrs Jean Todd Stephenson for her kind help with the English text.

Contributors

Ned Block, Department of Philosophy, MIT.
Paul Boghossian, Department of Philosophy, New York University.
Josep Corbí, Departamento de Filosofía, Universidad de Valencia.
Jerry Fodor, Department of Philosophy, Rutgers University and Ph. D. Program in Philosophy, The Graduate School and University Center of CUNY.
Richard Foley, Department of Philosophy, Rutgers University.
Alvin Goldman, Department of Philosophy, University of Arizona.
Paul Horwich, Department of Philosophy, MIT.
Richard Jeffrey, Department of Philosophy, Princeton University.
Jaegwon Kim, Department of Philosophy, Brown University.
Ernest LePore, Department of Philosophy, Rutgers University.
Jorge Rodríguez Marqueze, Facultad de Filosofía y Ciencias de la Educación, Universidad de Oviedo.
Stephen Schiffer, Ph. D. Program in Philosophy, The Graduate School and University Center of CUNY.
Ernest Sosa, Department of Philosophy, Brown University.
Stephen Stich, Department of Philosophy, Rutgers University.
Javier Rodríguez Alcázar, Departamento de Filosofía, Universidad de Granada.

Enrique Villanu
the Instituto de
cas in the Unive
ma de México
President of the
Americana. He
y Privacidad, 1
Filosófica, 198
and is the auth
coming. He is t
posio Internac
ed by the Un
noma de Méxi
the *Interdisci*
in Cognitive S
Nacional Autó
also Director
gram, *Consciousness, Semantic Representations and Reasoning*, supported by the Consejo Nacional de Ciencia y Tecnología, México.